The Sundarbans

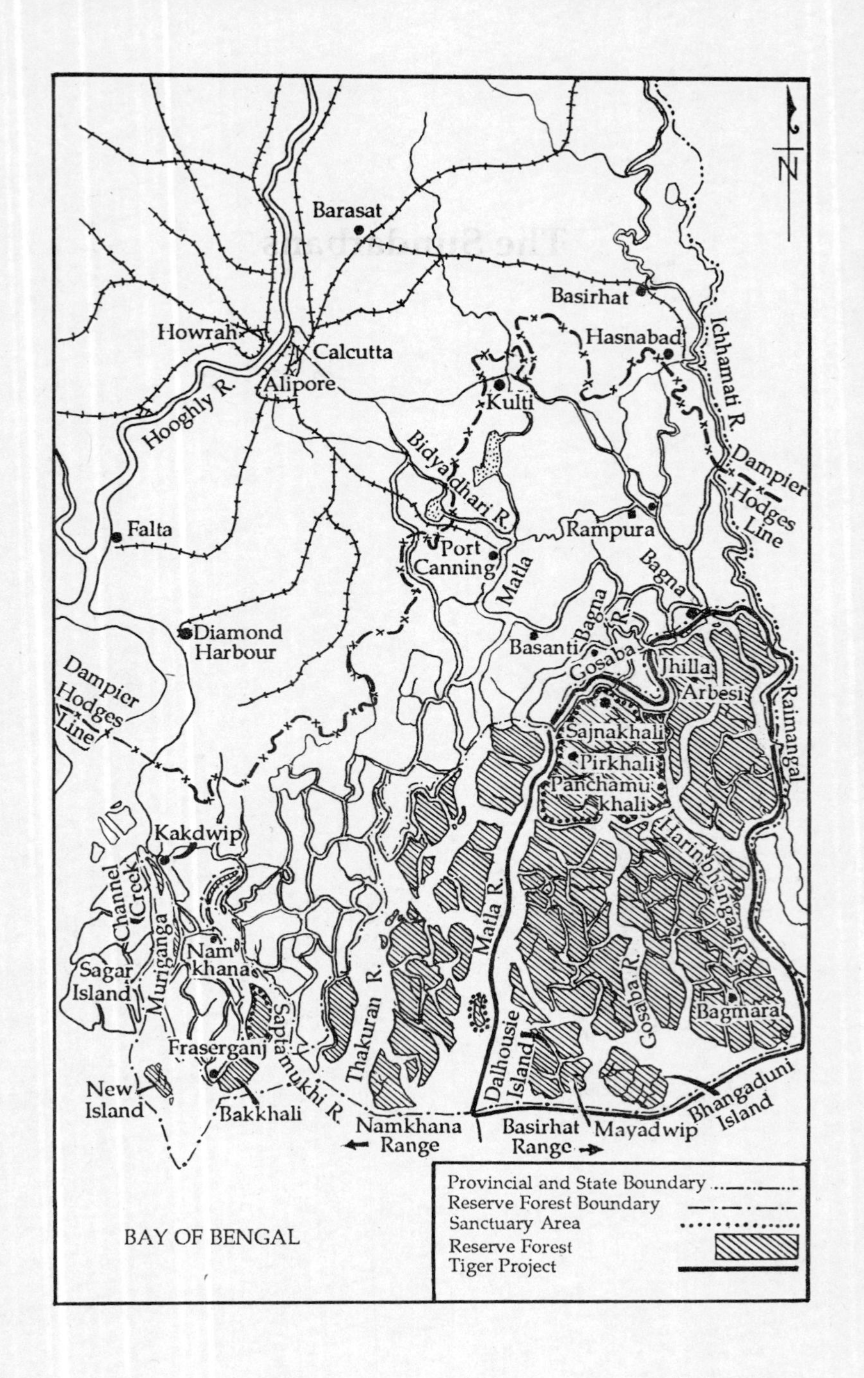

N
Barasat
Basirhat
Howrah
Calcutta
Hasnabad
Alipore
Kulti
Hooghly R.
Ichhamati R.
Bidyadhari R.
Dampier Hodges Line
Falta
Rampura
Port Canning
Matla
Bagna
Bagna R.
Diamond Harbour
Basanti
Gosaba
Jhilla
Arbesi
Raimangal
Dampier Hodges Line
Sajnakhali
Pirkhali
Panchamukhali
Harinbhanga R.
Kakdwip
Channel Creek
Muriganga
Namkhana
Matla R.
Thakuran R.
Sagar Island
Saptamukhi R.
Gosaba R.
Bagmara
Dalhousie Island
Fraserganj
New Island
Bakkhali
Namkhana Range
Basirhat Range
Mayadwip
Bhangaduni Island
Provincial and State Boundary
Reserve Forest Boundary
Sanctuary Area
Reserve Forest
Tiger Project
BAY OF BENGAL

The Sundarbans

Rathindranath De

CALCUTTA
OXFORD UNIVERSITY PRESS
DELHI BOMBAY MADRAS
1990

Oxford University Press, Walton Street, Oxford OX2 6DP
New York Toronto
Delhi Bombay Calcutta Madras Karachi
Petaling Jaya Singapore Hong Kong Tokyo
Nairobi Dar Es Salaam
Melbourne Auckland
and associates in
Berlin Ibadan

Printed in India by offset
at Graphitech India Ltd, Bidhan Nagar, Calcutta 700 091
and published by S.K. Mookerjee, Oxford University Press,
Faraday House, P-17 Mission Row Extension, Calcutta 700 013

To
Our Dedicated Forest Personnel working
under trying circumstances

Preface

As a district officer of North 24 Parganas I was acquainted with the Sundarbans for the first time in 1970. I was fascinated by its estuaries, wildlife and people. In 1982– 86 as the West Bengal Forest and Tourism Secretary, I visited the region frequently and developed a passion for it. On one visit, with a view to study wildlife at night, my family and I stayed overnight on the Haldi watch tower. On another visit, my friend and colleague Mr Dilip Bhattacharya felt that there was no good book on the Sundarbans and suggested that I write on the subject. The current volume is the culmination of the efforts I made on the suggestion. I am indebted to the former Sundarbans Field Director, Mr Pranabesh Sanyal, who has helped me immensely with scientific publications, photographs and other information. I must also thank my daughter Madhumita for her assistance in checking the scientific names of animals and birds.

I hope the present volume will be useful to the people interested in conservation of nature.

25 April 1990
Calcutta

RATHINDRANATH DE

Contents

1. Introduction

Landscape of the Sundarbans — Origin of name — The Sundarbans Tiger Reserve — The Sundarbans National Park — The Sundarbans Biosphere Reserve — Objectives — World Heritage List

Francois Bernier, who travelled in the Sundarbans in 1665-66, wrote : 'Among these islands, it is in many places dangerous to land and great care must be had that the boat, which during the night is fastened to a tree, be kept at some distance from the shore, for it constantly happens that some person or another falls prey to tigers. These ferocious animals are very apt, it is said, to enter into the boat itself while the people are asleep, and to carry away some victim, who, if we are to believe the boatmen of the country, generally happens to be the stoutest and fattest of the party.' The Sundarbans has been well-known for many centuries for its wild-life, especially the man-eating tiger. The problems such tigers create, are now being tackled on scientific lines.

The Sundarbans Tiger Reserve has the largest tiger population in India. Here the Bengal tiger swims as well as any aquatic animal and is recorded to have swum upto eight kms. at a stretch. The big cat drinks mainly the saline water of the tidal rivers and fights the largest crocodiles on earth during spring tides that spare hardly any land from inundation. Walking on land is impossible on account of the dense jungle and prong-like pneumatophores (breathing roots) of trees that pierce upwards out of the ground. For a visitor every moment of his journey through the vast, watery wilderness is an exploration of the mystic forests. The shimmering tidal waters bordered by mangrove trees are like a world of fantasy. The visitor forgets normal time-schedule and satisfies himself with a slow, lazy cruise against the tide along estuaries. The Sundarbans is captivating and anyone who has visited the area once, will surely love to visit it again.

Various explanations have been given about the origin of the name 'Sundarbans'. The name might have been derived from *sundar* and *ban* — a beautiful forest — or from *samudra-ban* or its distortion, *samuda-ban*, meaning the forest near the sea. However, it is now generally accepted that the name has come from *sundari-ban* or the forest of *sundari (Heritiera fomes)* trees, the species commonest in this forest. The name, Sundarbans, seems to be of recent origin. The Mughal historians used the name *Bhati* for the coastal strip between the Midnapore district and the Meghna river in Bangladesh, signifying low land subjected to tides.

Tiger hunting for sport and trophy increased all over the country since the turn of the century. Kailash Sankhala in his book, *Tiger!* writes that 86 tigers were shot in the Sundarbans alone in 1914. Speaking of the situation in the early seventies, he adds, 'The Indian Government then stepped in and conducted a formal census, and the figures we arrived at were not very different, only about 2000 wild tigers left, compared with some 30,000 within our own lifetime.' During his visits to the Sundarbans, Sankhala observed fresh pug marks of tigers in every part of the forest.

The Sundarbans is one of the tiger reserves in which Project Tiger was launched in 1973 to save the tiger from extinction. The Sundarbans Tiger Reserve covering 2585 sq. kms., had 135 tigers at that time and has 264 (1984 census) at present, the annual rate of growth being 8.7 per cent. The remarkable success of the Project in the Sundarbans in protecting the Bengal tiger as also the rare and beautiful mangrove ecosystem has been acclaimed by wildlife-lovers all over the world.

Outside the tiger reserve, there are two more sanctuaries at Halliday Island and Lothian Island.

The core area of the reserve covering 1,330 sq. kms. has earned the distinction of being the only national park in West Bengal and one of the 67 national parks in India, the oldest being the Corbett National Park in Uttar Pradesh. The Sundarbans National Park is bounded by the Harinbhanga river on the east, the Bay of Bengal on the south, the river Matla on the west and the forest blocks of Netidhopani, Chamta, Chandkhali and Bagmara on the north.

Incidentally, the first national park in the world, the Yellowstone National Park was set up in 1872 in the United States. A national park has the following features : (a) The basic objective is to conserve objects having aesthetic, geological, pre-historic, historical, archaeological or other scientific value for the enjoyment of the public. (b) No animal or plant is allowed to be hunted, killed or trapped in a national park but specimens of plants and animals can be collected for scientific studies with the permission of the park authorities. (c) National parks are managed by governments and their areas can be changed only under the due process of law. In the United States there is a separate enactment for the national parks. [In India the system of national parks is operated under the Wildlife (Protection) Act, 1972.] Africa has a number of national parks which are mainly sanctuaries for the wildlife, the importance of their scenic beauty being secondary.

Recently, the entire area of the Sundarbans, south of the Dampier-Hodges Line covering 9,630 sq. kms., has been declared a biosphere reserve. The Sundarbans Biosphere Reserve is the fifth biosphere reserve in the country, the other four being Nilgiri, Nanda Devi, Namdhapa (Arunachal) and Nakrek (Meghalaya). On the basis of Udvardy's modern bio–geographic classification seven more reserves are in the offing in the country. They are Uttar Khand, Gulf of Mannar (Tamil Nadu), Thar Desert, North Island of the Andamans, Manas, Kaziranga and Kanha.

The Sundarbans is the only mangrove tigerland in the world where the tiger occupies the pinnacle of both aquatic and terrestrial foodweb. As the Bengal basin suffered an easterly tilt on account of a neotectonic movement from the twelfth to the sixteenth century and the river Ganga which used to flow through the Indian Sundarbans, ran along the Padma channel on the east into Bangladesh, the mangrove tract within India now exhibits the special ecological effects of the tidal waters of the sea undeterred by any flow of fresh water. Thus the Sundarbans has a unique biogeographic climate. These are some of the considerations for which the Sundarbans has been selected as a biosphere reserve.

It may be mentioned that the Man and the Biosphere (MAB) Programme, launched in 1971, is a world-wide programme dealing with the people-environment interactions in the entire range of bioclimatic and geographic situations of the biosphere. The main theme of the programme is 'conservation of natural areas and the genetic material they contain'. The first biosphere reserves were designated in 1976 and since then the network has grown steadily, there being 243 reserves in 65 countries at present. The main characteristics of biosphere reserves are the following: (a) They are protected areas of representative terrestrial and coastal environments which have been internationally recognized for their value in conservation and in providing scientific knowledge, skill and human resources to support sustainable development. (b) They are united to form a global network which facilitate sharing of information about the conservation and management of natural and managed ecosystems.

The Sundarbans has earned yet another distinction — the entire forest area has been included in the World Heritage List of International Union for Conservation of Nature and Natural Resources.

2. Physical Aspects

Rivers in northern part — Estuaries and islands — *Khals* (canals) — Geographical divisions — Depositional history — 'Swatch of no ground'

The Sundarbans is a network of tidal rivers, creeks and islands. The seaboard area is a typical example of deltaic formation and exhibits the process of land-making in an unfinished state. It presents the last stage in the life of a great river emerging through a region of half land, half water, almost imperceptibly, into the sea. W.W. Hunter has described it as ' a sort of drowned land, broken up by swamps, intersected by a thousand river channels and maritime backwaters, but gradually dotted, as the traveller recedes from the seaboard, with clearings and patches of rice land.'

The principal rivers in the northern part of the Sundarbans are the Hooghly and the Bidyadhari. It may be noted that the main current of the Ganga has long been deflected to the east along the Padma into Bangladesh while its connection with the Hooghly has been silted up. However, a large volume of water is conveyed to the Hooghly during the flood season by the three Gangetic distributaries, viz., the Bhagirathi, the Jalangi and the Mathabhanga. In the dry season the Hooghly is largely fed by percolation and the tides running from the Bay of Bengal. It also receives some fresh water from the barrage over the Ganga at Farakka. Downstream of Calcutta the river receives the Damodar and the Rupnarayan which deflect the stream to the east. Below the town of Diamond Harbour, the river resumes a southerly course until it debouches in the Bay of Bengal. Shortly before its confluence, it bifurcates, the main channel passing west and another channel east of Sagar Island. The latter channel is called the Channel Creek, locally known as the Muriganga. On the west bank the main channel receives the Haldi and the Rasulpur rivers. The greatest mean rise of tide in the Hooghly, taking place in March-May, is about five metres, the minimum being one metre. The tide is occasionally so strong that it causes the phenomenon known as a bore. A bore is nothing but a headwave formed when an unusually high tide is checked by the narrowing of the river channel, the height of the wave often being two metres or more. An account in *The Calcutta Review* of 1859 testifies to its danger : 'Upon the approach of this wave a distant murmur is heard, which turns into the cry, *ban ! ban ! ban !* (tidal bore) from the mouths of thousands of people, boatmen, sailors and others, who are always on the look-out for this much dreaded wave. This cry is the signal for all sorts of craft to push out into the centre of the river, the only spot where the wave does not curl over and break. Should any boat or larger craft be caught in that portion of wave that breaks, instant destruction is inevitable.' Navigation in the Hooghly has become difficult on account of rapid currents, shoals and shifting sandbanks. However, with some dredging operations the river has been maintained navigable to large liners and is the gateway for the foreign trade of eastern India.

The present channel of the Hooghly is very different from the original which has been identified with Tolly's Nullah from Kidderpore to Garia (13 km. south of Calcutta), from which point it ran to the sea in a south-easterly direction. It is said that it emerged out of the Sundarbans at Kakdwip and passed along the present Muriganga, after which it flowed through a creek in a westerly and then in a southerly direction until it fell into the Bay of Bengal at Ganga Sagar. The original course is still traceable to some distance beyond Garia, being known as *Adi* Ganga (the original Ganga), *Buraha* Ganga (the old Ganga) and Ganga *Nullah* (the Ganga canal).

The Bidyadhari, a tidal river with a circuitous course, begins in the Sundarbans. It flows north-east past Haroa as the Haroa Gang. Then it flows south-west to the junction of the Beliaghata Canal and Tolly's Nullah and thence south-east to Canning where it is joined by two other rivers. The united stream forms the Matla river which flows to the sea and is navigable by river steamers upto Canning. The stretch of the Bidyadhari near Calcutta, which serves as an outfall channel for the storm-water and sewage of the city, has silted up on account of the activities of local fisheries and the reclamation of portions of the Salt Water Lakes for habitation.

The Kalindi forms the eastern boundary of the Indian Sundarbans down to the sea where it merges in the Raimangal estuary.

The south of the Sundarbans is intersected by enormous tidal rivers, some of them being of great size, formed by the joining of smaller water courses and branches thrown off by other rivers, all having a general southerly course towards the sea. The most important ones from west to east are the Channel Creek, the Saptamukhi, the Thakuran, the Matla, the Gosaba and the Harinbhanga.

The main estuaries or arms of the sea, from west to east are the Channel Creek, the Saptamukhi, the Thakuran, the Matla, the Gosaba and the Raimangal, the last including the mouths of the Harinbhanga and the Kalindi. The tidal fluctuations in the rivers and estuaries are remarkably high, ranging from nearly six metres to less than one.

The estuaries are separated by large islands, among which the main ones are Sagar, Fraserganj, Lothian, Halliday, Dalhousie and Bhangadhuni islands.

Between the large estuaries and rivers there are innumerable streams and water courses, called *khals*, forming a perfect network of channels and ending in little channels which draw off the water from every block of land. Each block is like a saucer with high ground along the bank of the *khals* surrounding it and with one or more depressions in the middle. The water collects in the depressions and is drained off by the little *khals* into the larger khals and ultimately into the rivers. Many of the *khals* connect two larger ones and as a result the tide flows into them through both ends. Such *khals* are called *do-aniya khals*.. They afford communication between the larger *khals*, but they get silted up at the point where the two tides meet because the water is stagnant there.

The inhabited part of the Sunderbans has some marshes and swamps (*bils*). Large areas of marshland have been brought under cultivation by means of embankments raised to keep out brackish water.

The whole of the Sundarbans may be divided into three distinct geographical regions. Along the Hooghly the land is high, but east of it, as far as the river Kalindi, i.e. in the 24-Parganas district, the low-lying land is protected from submersion by immense embankments. From the Kalindi to the Baleswar, i.e. in Khulna (Bangladesh), the land being moderately high, the embankment is only a few feet high. Further east i.e. in Bakerganj (Bangladesh), the land being high no embankment is necessary. Depending on these conditions, the landscape also changes in a striking way. Bakerganj looks prosperous and every well-off farmer has a good homestead with many fruit-trees and some winter crops in addition to rice, which is practically the only crop that can be grown in the Sundarbans. In the region of the 24-Parganas, fruit-trees do not thrive and the area wears a less cheerful look. Khulna, having intermediate conditions, gives a corresponding look. However, even there, large tracts are so swampy that the people who cultivate them are obliged to live elsewhere.

There are good reasons to believe that there has been some subsidence of the country followed by deposition in comparatively recent times. *The Calcutta Review*, July 1889 has recorded : Stumps of *sundari* trees were found embedded near Sealdah in Calcutta thirty feet below the ground. It is known that the *sundari* does not grow except between high-water and low-water levels. So, the trees could not have grown naturally as they were found.

The depositional history of the region has been emphasized by the findings thrown up by the excavations of the Metro Railway Project in Calcutta, the Kolaghat Thermal Power Project and the Second Hooghly Bridge Project. The distinct deposition of a layer of 0.50 to 1.20 m. thick peat (dark brown humus formed by the partially decomposed mass of marsh-vegetation) was recorded five metres below the surface, being about the mean sea level. In the excavations of the Metro Railway Project a second layer of peat has been recorded for the first time 12 metres below the surface, being seven metres below the mean sea level. The sequence of sediments with two distinct layers of peat and an intermediate layer of soft grey clay with decayed stumps of *sundari* trees has been found at Jatin Das Park, Bhowanipur and Esplanade stations of Metro Railway. The remains of animals, like the skull and vertebrae of gharial (*Gavialis gangeticus*) and the carapace of the chitra turtle (*Chitra indica*), both fresh water loving animals, have also been recorded in some recent excavations. On the basis of these records, scientists have suggested that the southern part of the Bengal basin had five phases of vegetation. The phases included tidal mangrove, followed by sudden cessation of tidal influx, marine, and salt tolerant fresh water vegetation. This suggestion supports the view of the eastward tilting of the basin some 300 years ago.

There is a great natural depression called "Swatch of no ground" in the Bay of Bengal, south of the Raimangal estuary and this figures in the navigational charts. The surrounding waters which are around 20 metres, change almost suddenly to 500 metres in depth. Fergusson's theory is that the sediment is carried away from the spot and deposition prevented by the strong currents produced by a meeting of the tides from the east and west coasts of the Bay of Bengal. Another theory attributes the depression to a local sinking. However the mystery shrouding the origin of the 'Swatch of no ground' has by no means been cleared up.

3. History

Barah Bhuiyas — Pratapaditya — English – Mughal conflict — Portuguese and *Mag* pirates

The Mahabharata and the Puranas contain references to this part of the Gangetic delta. They indicate that this area lay between the kingdom of the Suhmas in western Bengal and of the Vangas in eastern Bengal. The Suhmas lived near the sea-coast on a great river which has been identified by some historians with the Bhagirathi. At the time of the Raghuvansa of Kalidasa the country appears to have been subject to the Vangas and the epic mentions the defeat of the naval forces of the Vangas by Raghu who 'establised pillars of victory on the isles in the midst of the Ganges'. In all probability, these islands represented the present area of the 24-Parganas which was still a low marshy tract of land intercepted by rivers. Ptolemy's map of the second century shows the south of the delta so much cut up by rivers and estuaries that it was practically a collection of islands.

However, nothing concrete is known of the area until the end of the fifteenth century when a few details may be gathered from a Bengali poem of Bipradasa and from the *Ain-i-Akbari*. The poem of Bipradasa, describing the voyage of a merchant called Chand Saudagar from Burdwan to the sea, mentions that the merchant passed Calcutta and proceeded along the Adi Ganga. The *Ain-i-Akbari* embodies the rent-roll drawn up by Todar Mal in 1582. The rent-roll of Todar Mal also refers to Calcutta (Kalikata) as a *mahal* (revenue unit).

The actual ruler of the Sundarbans towards the end of the sixteenth century was a Hindu chief called Pratapaditya, one of a group of chiefs known as *Barah Bhuiyas* (twelve chiefs). The twelve chiefs were nominally vassals of the Delhi emperor, but really enjoyed independence in the south and south-east of the Gangetic delta. The Delhi emperor, Akbar's armies were engaged in suppressing the military revolt and campaigns against the turbulent

Afghans who had made themselve masters of Orissa and part of Bengal even after the death of Daud. Consequently the *Barah Bhuiyas*, secure in their swampy retreats, paid no tributes and displayed a royal splendour. They did not, however, call themselves kings.

Pratapaditya is regarded by the Bengalis as a national hero. Legends have it that his father, Bikramaditya made his capital at Iswaripur, now a small village in the Khulna district (Bangladesh), 20 kms. south of Kaligunj. The place was also known as Jasohara or Jasor. Pratapaditya moved the capital to Dhumghat, another place in the Sundarbans, the actual site of which is doubtful, but which could not have been far from Iswaripur. He extended the limits of his kingdom by conquests till the surrounding country acknowledged his rule. He declared himself independent of the Mughal emperor and defeated the commander-generals sent against him. In course of time he became a tyrant and killed his uncle, Raja Basanta Roy, who had showered much affection on him. An army under the command of Man Singh, Governor of Bengal, marched against him. He was taken a prisoner and his capital was captured. Pratapaditya, preferring death to dishonour, poisoned himself to death.

Some historians have identified Pratapaditya with the king of Chandecan. This king finds mention in the letters of Jesuit missionaries who visited Bengal at the end of the sixteenth century. The two priests, Fernandez and Josa, who arrived at Hooghly in 1598, were invited by the king of Chandecan to pay him a visit. Fernandez's account of their journey shows that the route lay in the Sundarbans, the capital being at a place situated half way between Chittagong and Hooghly. The king's dominions were so extensive that it took 15 to 20 days to traverse them. The two priests encountered great dangers on the way both from dacoits and from tigers. The country had a great trade in bees' wax which was produced in the jungles. Around this time a church was built at Chandecan and this was the first Christian church erected in Bengal. In 1602, Carvalho, the Portuguese commander of Sandwip Island and some Jesuit priests were summoned by the king to Jasor. Though the king promised friendly conduct to them, Carvalho was put to death and the priests were driven out.

Chandecan has been identified by H. Beveridge with Pratapaditya's capital of Dhumghat which is placed in the neighbourhood of Kaligunj in Khulna. Another theory is that Chandecan was Sagar Island.

It is believed that Pratapaditya established a naval centre at Sagar Island where a number of ships were always ready for battle. The three other places where Pratapaditya built his shipyards were Dudhali, Jahajaghata and Chakrasi. It is said that Rodda, the Portuguese Admiral of Pratapaditya, defeated the Mughal forces in a battle at the confluence of the Adi Ganga and the Bidyadhari.

In the latter half of the sixteenth century the emporium for the sea-borne trade of Bengal was Satgaon on the Saraswati river near Hooghly. As the Saraswati silted up, Satgaon was gradually superseded by Hooghly as a commercial centre and some of its inhabitants migrated to other places. Four Basak families and one Seth family founded the village of Govindapur on the site of the present Fort William in Calcutta. Another market was also established at Sutanati, where present-day north Calcutta is situated. The new settlement of Sutanati attracted Armenian merchants also. It was here that the English under Job Charnock sought refuge after abandoning the factory at Hooghly in 1686.

There was friction between the English and the Mughals owing to the latter's extortions and interference with the Company's trade. Ultimately in 1690 the Mughal emperor granted the English licence for trade. On Sunday, August 24, 1690 at noon the weather-bitten band of Charnock, anchored, for the third time, near the muddy banks of what was to become the British capital of India. The miseries of the fever-stricken band in the next two years are beyond description. The only place on which the pioneer settlers could build their houses was a narrow strip of land on the river bank. The Salt Water Lake on the east left masses of dead, putrid fish as the water receded in the dry season while a swampy jungle ran up to where Government House now stands. Popularly, Charnock's choice of Sutanuti is ascribed to chance. The story is that Charnock was delighted with the charms of the place while smoking a *hookah* in the shade of a large peepul tree near the present Sealdah station or of a *nim* tree near the present Nimtala ghat.

The Portuguese, who occupied Tardaha on the Bidyadhari towards the end of the sixteenth century, combined piracy with trade. The whole of the Sundarbans was infested with Portuguese and *Mag* pirates. From this situation the Channel Creek earned the sobriquet, Rogues' River, in the eighteenth century. It is believed that the fort of Tanna was built at the site of the present Botanical Gardens to check incursions by pirates and a chain had to be run across the Hooghly between Calcutta and Sibpur to prevent pirates extending their raids up the river. *The East India Chronicle* for 1758 records that the *Mags*, in February 1717, carried off from the southern parts of Bengal not less than 1800 persons — men, women and children. They were taken to Arakan in Burma where the king chose the artisans, about one-fourth of those captured to be his slaves while the rest were sold as slaves at prices ranging from Rs.20 to Rs. 70. Certain remarks of Sir William Jones, Chief Judge of the Supreme Court, indicate that slavery was also practised in Calcutta towards the end of the eighteenth century.

4. Revenue History

Reclamation of forests — Dampier-Hodges Line — Reserved forests

At the time of the East India Company's acquisition of the civil administration of Bengal in 1765, the forests of the Sundarbans extended to the vicinity of Calcutta. The process of reclaiming the forests started with the idea of converting that dreadful region of smugglers, pirates and wild animals into revenue-yielding holdings. The history of this reclamation makes interesting reading.

The first effort to reclaim the forests was made in 1770 by Claude Russel, Collector-General, 24-Parganas, who allowed the lessee an initial period free of rent. Substantial progress was made by the lessees and the neighbouring zamindars also busied themselves in reclaiming forests. During the next 40 years the country was cleared almost down to Sagar Island on the south and nearly as far as Port Canning on the east.

The next effort was made in 1783 by Tilman Henckell, Judge and Magistrate of Jessore, who decided to lease out small plots directly to ryots. The scheme was opposed by the zamindars who claimed the lands cleared by the grantees. In 1786, Henckell demarcated with bamboo posts what he considered to be the northern boundary of the forests, but notwithstanding all his efforts, the struggle with the zamindars was too severe for the lessees, most of whom had disappeared by 1792. In the cases of those who remained, the character of the scheme had been modified and the lessees developed into *talukdars* and their lands were called 'Henckell's *taluks*'.

It may be mentioned that the exclusion of these leases and forests from the Permanent Settlement has rendered the history of the Sundarbans peculiar. Around 1810, various schemes, like the reclamation of Sagar Island and construction of wet-docks at Diamond Harbour, drew attention to the Sundarbans. The Sundarbans excluding the sea face, from the Hooghly to the river Passur, was surveyed by Lieutenant W.E. Morrieson in 1811-14 and his results were corrected by his brother, Captain Hugh Morrieson in 1818. This was a great work and has been the basis of all subsequent maps of the region. In course of his survey Lieutenant Morrieson found that the north-east branch of the Raimangal estuary was within a small distance of the Kalindi. He made a cut joining the two rivers. The stream of the latter soon enlarged the cut and a large quantity of its fresh water was diverted into the Raimangal. At that time cultivation extended much farther south on the east bank than on the west bank of the Kalindi, but the diversion of fresh water deprived that country of its chief advantage and a considerable tract reverted into jungle. This is an instance of far-reaching disastrous consequences arising out of a seemingly trifling act.

During 1814-16 an attempt was made to remeasure the grants already made and revise their rentals. By that time, the advantages of the opening up of the Sundarbans were clearly perceived and a law sanctioning the post of Commissioner in the Sundarbans was passed in 1816.

The first Commissioner, D. Scott, began surveying the grants, south and east of Calcutta and found that encroachment and reclamation had been continuously progressing, partly by the lessees, partly by the zamindars and partly by unauthorized persons. The newly reclaimed area was held without payment of any revenue to the state. The proposal to levy revenue was opposed by the zamindars who claimed the whole forests and resisted the survey tooth and nail. Indeed, Scott had to be allowed an escort of 12 sepoys for his protection.

Ensign Prinsep surveyed the line of dense forests from the river Jamuna to the Hooghly in 1822-23. With the help of the Morrieson's map, he divided the forests lands concerned into blocks and numbered them. This was the beginning of the 'Sundarbans lots'.

William Dampier, Commissioner, and Lieutenant Hodges, Surveyor for the Sundarbans, defined and surveyed the line of dense forests from the Jamuna to the eastern limit in 1829-30. In 1832-33 Dampier formally affirmed Prinsep's line in the 24-Parganas. 'Prinsep's Line' and 'Hodges' Line' are the authoritative limits of the Sundarbans forests. Hodges prepared a map of the whole Sundarbans which has been the standard map ever since.

In accordance with the rules of 1830 for the grant of forest lands, applications poured in mostly from the European residents of Calcutta. Applicants whose number included some local gentlemen like Radha Krishna Dutta, Guru Prasad Chowdhury and

Hafizuddin, practically got whatever grants they asked for in the 24-Parganas and Khulna. These grants were made in perpetuity at a rental of Rs. $1^1/_2$ per acre and nothing was payable within 20 years, but one-fourth of the area was to be reclaimed within five years.

F.E. Pargiter has detailed the procedure of reclamation in *The Calcutta Review* of July 1889 : The first thing was to embank the lands. This meant that a line was cut through the forest along the banks of the streams surrounding the lot, and an embankment thrown up along the line; and that strong dams were constructed across the mouths of the smaller streams running into the block in order to keep the salt water out. When this work was finished, the work of felling the forest, digging tanks, and constructing huts for the future cultivators began. All these operations required months of constant attention. A strong body of coolies was procured who were supported all the while with food and fresh water. Their health and safety was cared for, and a *shikari* (hunter) was generally employed to fire his gun occasionally to frighten away the tigers that abounded in the forest; if there was no *shikari*, the coolies raised a combined shout at short intervals. Unusual sickness or destruction by tigers produced a panic and inflicted almost irreparable loss.

The grantees found it difficult to collect the labourers they wanted or a sufficient number of ryots to settle on the lands permanently. The labourers were the Sundarbans woodcutters, coolies from Chota Nagpur and *Mag* coolies from the eastern frontier. The difficulties of the grantees whose lands lay deep in the forest were greater; but even those who got grants bordering on cultivated tracts and were in a comparatively advantageous position, suffered severely from the hostility of the zamindars. To add to all these difficulties, great damage was inflicted on the grants by the periodic cyclones and waves. Some of the grantees, being speculators, did not attempt to clear their lands, but realized whatever profit they could get from the wood and other natural products and sold the lands as soon as they found a purchaser.

The rules of 1853 stipulated that grants for 99 years were to be made to the highest bidders. The revenue was reduced to six annas per acre. Reclamation was more carefully provided for and the grantee was required to reclaim one-eighth of his land in five years, one-fourth in 10, one-half in 20 and the whole in 30 years, under pain of forfeiture.

The Waste Land Rules, 1863, provided for exemption from revenue in cases of outright sale and redemption of land revenue by payment of a capitalized sum.. However, many purchasers were unable to pay the purchase-money even by instalments. The rule for the redemption of revenue met with more success.

In 1862, Dr. Brandis, Conservator of Forests in Burma urged the Government to conserve the forests of Bengal. In 1878-79, 4,879 sq. kms. of forests were declared protected, but leasing out for cultivation continued. Between 1928 and 1943 the forests were declared reserved.

In 1865, F. Schiller and eight other gentlemen, including some Europeans, applied to the government to purchase all the remaining waste-lands, proposing to raise a capital of not less than one million sterling and to reclaim the lands by importing labour from China, Madras, Zanzibar and other places. The government declined to approve any import of labour from Africa. Schiller failed to induce the public to join his company. So much money had been lost in reclaiming the Sundarbans that people had grown cautious.

The rules of 1879 provided for two kinds of grants, viz. (a) blocks not exceeding 200 acres leased to small settlers for 30 years, lands to be reclaimed within two years, and (b) blocks of 200 acres or more leased to large capitalists for 40 years, one-fourth of the area being exempted from assessment in perpetuity and one-eighth to be reclaimed within five years. Experience showed that in the capitalist system the actual cultivators were oppressed. In 1904 the capitalist system was abandoned and a ryotwari settlement was introduced. This settlement was tried on a large scale in Fraserganj. The reclamation proved very expensive and cultivators could not be settled on remunerative terms.

On the whole, in most of the Sundarbans grants a revenue-free period was given and there was a stipulation that within a certain period a certain area must be reclaimed. This put a great responsibility on the lessees who were anxious to shift some of it. The lessess divided their area into several portions and leased them out to persons who would undertake the duty and cost of reclamation and induction of tenants. In a few cases only the lessee himself inducted tenants while in most case the lessees were absentees from the first date of the grant and took fixed payments from their tenure-holders.

5. Flora

Special features of mangroves — Natural regeneration — Normal pattern of estuarine vegetation — Abnormal adaptations of the Sundarbans mangroves

The swampy islands are covered with dense forests of mangrove which is especially suited to the fine clayey soil, the salinity of the water and the tidal current which continually brings in and deposits fresh soil. The most striking feature of some species is, perhaps, the stilt roots which both support the plants and are their means of respiration. The *garjan* (*Rhizophora apiculata*), *genwa* (*Excoecaria agollocha*) and *jele garan* (*Ceriaps decandra*) are examples of such plants. Another noticeable feature is the pneumatophores (breathing roots) of plants such as the *kala baen* (*Avicennia marina*), *keora* (*Sonneratia apetala*), *sundari* and *passur* (*Xylocarpus mekongensis*). Plants such as the *khalsi* (*Aegeceras carniculatum*), *tora* (*Aegialitis rotundifolia*) and *kala baen* have salt-excretory glands. The cell sap develops high osmotic pressure which helps the plants draw water from the concentrated soil solution. All these plants have viviparous germination. The *dhanighash* (*Porterasia coarctata*) which grows on fresh deposits of silt alongside the forests or on freshly-formed island in the midst of rivers, is the dominant grass species in the region. Herbaceous growth is strangely absent. Plants of *genwa* and *jele garan* species account for about 70 per cent of the plants. *Keora* is the tallest plant in the region and its sour fruits make delicious *chatni* (jelly-like preparation).

Incidentally, forests in the Sundarbans develop under the process of natural regeneration. With adequate protection from human interference freshly-formed lands get covered with mangrove plants within a few years. The planting of seedlings which involves a lot of labour and expense, is not required in the region.

During the past 600 years the Ganga basin has gradually tilted towards the east and the tributaries — the Matla and the Bidyadhari — have gradually got detached from the main sweet water source. As a result, except for the Hooghly river which still receives some fresh water from the Farakka barrage, the rivers, channels and creeks in the Indian Sundarbans are now filled with only tidal waters. This has resulted in abnormal salinity in the different zones of the estuaries. The absence of a flow of fresh water invites more deposition by the backwaters of the sea and the result is accelerated geomorphic action. Recent studies indicate that this has given rise to abnormal adaptation of mangroves in the region.

In the normal pattern of vegetation in an estuarine delta, distribution of mangrove communities on tidal flats depends on the tidal intensity, the type of sediment and the salinity. Three zones are generally recognized : (a) A true estuarine zone, comprising the estuarine banks along the mouths of rivers, is mainly dominated by the *kala baen, tora* and *kripa*. These species tolerate high salinity and submergence either by developing salt excreting glands or by increasing water storage tissues in their leaves, and develop soft and porous pneumatophores. (b) A middle estuarine zone is dominated by the *garjan, jele garan* and *keora*. Here the salinity is lower than in the true estuarine mouth, but the tidal current passing through narrow creeks and channels is higher. Mangroves adjust to these habitats by producing stilt roots; pneumatophores are soft and porous. (c) An inner estuarine or riverine zone comprising elevated areas with less aerated soil and more fresh water flow, is dominated by the *sundari, genwa, kankra* (*Bruguiera sexangula*) and *golepata* (*Nypa fruiticans*). These species usually prefer more fresh water and develop hard pneumatophores to adjust to less aerated soil.

Scientists have observed that this normal zonation, which is present on the west of the river Saptamukhi owing to the influence of the fresh water flowing down the Hooghly, is altered in the major portion of the Sundarbans, east of the Saptamukhi. In this area the dominant *kala baen* and *tora* communities of the true estuarine zone are progressively lost as erosion undercuts mangrove roots and the *garjan* and *jele garan* communities of the middle zone are prominently seen in these truncated zones. In some places the true estuarine zone is found to be lost on account of collapse along the cliff eroded area and the inner estuarine types, *sundari* and *golepata* become exposed along the estuarine mouth.

Over the years the stock of *sundari* trees in the region has got depleted. *Sundari*, being valuable timber, has been over-exploited in the past. Secondly, as already mentioned a great part of the inner estuarine zone has become more saline due to the lack of fresh water and the *sundari,* which prefers less saline soil, does not proliferate nowadays.

Of the forest products *golepata* is used as roofing material for rural houses; *sundari* and *passur* as also *dhundul, garjan, tora* and *mathgaran* are used for house posts. *Garan, tora, garjan, kankra* and *hental* (*Phoenix paludosa*) are used as rafters. Branches of *garan* are used for wattle. *Sundari* wood is good for making boats. The most favoured fuels are *garan, khalsi, baen* and *singara.*

6. Fauna

Food chain — Description of fauna — Tiger, jungle cat, civet cat —Axis deer — Deer-monkey relation — Wild pig — Water monitor — Estuarine crocodile — Shark and dolphin — Snakes — Horseshoe crab — Sajnekhali Bird Sanctuary — Other birds — Fish — Fishing at Jambudwip — Extinct animals

A special feature of the mangrove habitat is that its entire land mass gets submerged during a cyclone and the land animals have to play amphibious roles. The food-chain of the Sundarbans animals brings into focus the fact that the tiger is at the apex of the hierarchy of terrestrial as well as aquatic animals (See diagram below). Here, the tiger's prey includes pig, deer, monkey, water monitor, bird, crab and fish.

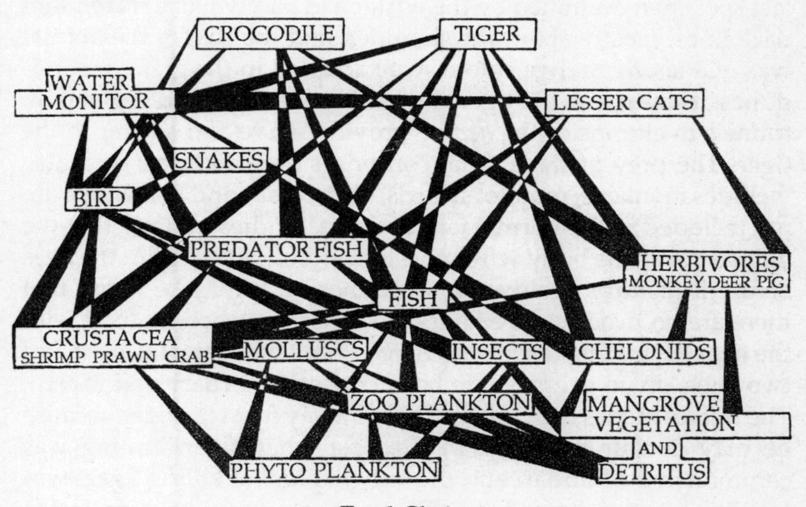

Food Chain

The magnificent Bengal tiger (*Panthera tigris*) — not Royal Bengal, E.P. Gee thinks the 'Royal' may have originated from the fact, a tiger was shot by the Duke of Windsor when he was the Prince of

Wales — holds the pride of place among the fauna of the Sundarbans. It is almost impossible to land anywhere in the forests without coming across the pug marks of the tiger. This is because the tiger reserve has as many as 264 tigers (1984 census) — 137 males, 112 females and the rest cubs — and the animal is a great wanderer covering long distances in search of prey and readily swimming across the rivers and *khals* that criss-cross the forests. The Sundarbans tiger's swimming capacity is proved by the report that a tiger was recently found in Halliday Island in the Matla river. In order to reach the island, which is generally free from tigers, the tiger must have swum 8 kms. at a stretch.

O'Malley's gazetteer has recorded : 'In their markings these animals vary greatly. The young tigers are handsomer than the old ones; their colouring is more vivid and the stripes darker and closer together. A curious adaptation to environment is seen in the tigers frequenting the sand-dunes of the sea face. These sand-dunes are covered with tall, brown spear-grass and immediately behind lie the glades of green herbage on which deer feed and pigs come out. A tiger, in such surroundings, would be rendered conspicuous by his stripes, so the sand-dune tiger has almost lost his stripes. The last specimen examined by the writer had barely half a dozen thin dark lines, mostly about the shoulder, and the coat of the animal was of a tawny orange colour, well adapted to the grass or sand-dunes. It was not an old tiger'. In the absence of ground vegetation, tunnels in clumps of the *hental* provide shade and shelter to the tiger. The prey of the animal comprises pigs and deer and also includes monkeys, monitor lizards, birds, crabs and fish which are not included in the normal fare of tigers. Studies indicate that the animal covers its body with mud to ward off attack from the bees before it disturbs beehives to drink honey. It may be noted that there are no major co-predators to share the sources of food with the tiger. Its mating season is between September and October and two cubs, on an average, are born between February and March. The tigress nurses her cubs and keeps away from the tiger because he may devour the cubs. In the past when tiger-hunting was common in the Sundarbans, the only method of killing tigers was by sitting over kills or baits. No elephants could be used in the forests on account of the pneumatophores and dense *hental* clumps.

The jungle cat (*Felis chaus*) and the civet cat (*Viverra zibatha*) are found in the forests near habitation. The fishing cat (*Felis viverina*) is distributed in the outer estuarine zone (within 20 kms. from the seaface) and the inner estuarine zone (beyond 60 kms. from the seaface); it is strangely absent in the mid-estuarine tract. The region has about 500 fishing cats.

The chital or axis deer (*Cervus axis*) is found in plenty — its population is estimated to be 30,000 in the Sundarbans. It feeds mainly on leaves, twigs and fruits of plants like *keora*. Its cry is a short high-pitched kind of bark, generally used as a signal or an alarm, the pitch of an alarm-call being greater than that of an ordinary call. During the rutting season, stags are seen engaged in long combats, pushing each other, with antlers interlocked, in open areas and emitting guttural sounds.

An interesting relationship between the axis deer and the rhesus monkey is observed in the Sundarbans. Hopping between keora trees monkeys drop fruits and green twigs which are eagerly browsed on by deer grazing in the shade. Monkeys, having excellent vision, warn the deer against a stalking tiger by raising alarm-calls from their position of vantage. Before the launching of Project Tiger the poaching of deer was a serious problem in the region. Many poachers used to deceive the deer into the range of their firearms by mimicking the call of monkeys from their positions on tree-tops.

The barking deer which was distributed in many areas of the eastern Sundarbans, is now restricted to Halliday Island in the Matla.

The wild pig (*Sus scrofa*), estimated to be 12,000 in number, is found all over the Sundarbans. The young of the animal is dark brown with longitudinal stripes of a lighter shade. These stripes disappear after a few months and the animal puts on a black coat. The boar is armed with long tusks which measure over 9 inches. Wild pigs go about in herds numbering four to ten. A visitor is likely to see the wild pig busily digging away at the roots of old trees and greeting him with a grunt, and a great rush of the unseen herd as it hurries into the dense forest. Pigs are less alert than deer. No wonder pigs provide 60 per cent of the food for tigers and deer only 20 per cent.

The otter (*Lutra nair*) is found in parts of the Sundarbans. The animal is tamed by fishermen and trained to drive fish into their nets.

Among lizards, the water monitor (*Varanus salvator*) and the yellow monitor (*Varanus flaviscense*) are the most prominent. The water monitor, the second largest monitor, is a rare species and needs to be protected. The animal attains a length of ten feet and is dark olive in colour. The tanks at Sudhanyakhali, Netidhopani and Haldi are the favourite haunts of this animal. During his visits to the Sundarbans, the writer has been thrilled by the sight of water monitors swimming in these tanks and later disappearing in the nearby forests. The water monitor is adept in climbing trees. The yellow monitor is about five feet long.

As a visitor sails down the tidal rivers he may be rewarded by the sight of an estuarine crocodile, with its ugly head, frightful jaws, lizard-shaped body and tough, hard skin, lying on mud-flats resembling a log of wood. The writer recommends a cruise between Netidhopani and Sudhanyakhali in the cold season for the purpose. The animal glides into the water and disappears with a splash soon after the launch nears it. The crocodile grows to a great length; a specimen in the British Museum is 33 feet long. The animal is an endangered species now as it used to be mercilessly hunted for its skin in the past. Many incidents testifying to the ferocity of the estuarine crocodile have been recorded in *The Calcutta Review* of March 1959 from the field diaries of the Morrieson brothers. One incident reads as follows : Observing a crocodile on the bank, the Morrieson brothers fired at it. It was wounded and after getting into the water, it again came to the bank. The Morriesons went in a boat and put two balls into its head. It charged open-mouthed at the boat, but sank from pain before it could reach the boat. When it rose about ten yards away, two balls were put through its body. The animal charged again and got in below the boat apparently attempting to upset it. But it failed and ultimately was found dead on the bank the next day. It was 15 feet in length, and on cutting open its belly bangles, rings, and other ornaments were found.

Sharks are found in the rivers and estuaries. They are 10 feet long and with their razor-sharp teeth can bite off the submerged part of human body without causing any pain to the person. In the past there were many incidents in which villagers lost their limbs to sharks while bathing in the rivers. Gangetic dolphins that are seven feet long, are also found in the region. During his visit to the Sundarbans the writer has enjoyed the sight of dolphins raising their heads above water to breathe and disappearing into the water again.

Of the snakes, the king cobra, common cobra, banded krait, Russel's viper and some sea-snakes are venomous while the python is the most prominent non-venomous snake found in the Sundarbans.

In addition to the edible red-crabs (*Scylla serrate*) there are 48 other varieties of crabs in the waters, the fiddler, hermit, tree-climbing, ghost (found on islands near the seaface) and mataplax crabs being some of them. It is interesting to note that hermit crabs live inside dead shells of snails. Oysters are also found in the water.

The horseshoe crab or 'living fossil', closely related to spiders and scorpions, is found in the region. The animal has been living on the earth for 600 million years. Though it dominated the sea once, now only three or four species are found in the North Atlantic ocean and far eastern waters. The adult life span of the crab is from five to seven years. It attains a length of two feet including a tail-like organ called the telson. The animal feeds on molluscs and worms dug up from the mud. Though it is capable of swimming, it is not particularly mobile. Its telson helps in burrowing and is also useful for righting a crab that has fallen on its back. Its mating takes place in the shallow coastal waters.

The Sajnekhali Bird Sanctuary is alive with birds during the monsoon. The sanctuary is a bird watcher's paradise during the period. The open bill stork, little egret, large egret, grey heron, purple heron, night heron and cormorant are the local birds coming from nearby areas while the most prominent migratory bird is the pelican. These varieties of birds start nesting on *baen* trees in the sanctuary between June and July and leave the sanctuary in September with their young ones. There are two watch-towers on the periphery of the sanctuary. From the Sajnekhali tourist lodge the place is 10 minutes away by country boat. Water monitors, being adept in climbing trees, kill young birds in their nests for food. They prove to be a danger to the sanctuary. Forest officials protect the young birds by tying thorns of *hental* to tree-trunks. Once the writer observed a couple of water monitors swimming in the tidal waters in the sanctuary, one of them holding a young bird in its mouth.

Numerous wading birds are found all over the region. Jungle fowls are also common. The shrill clarion call of the cock and the cackling of the hen with her brood of chickens, sounds common in a village, sound strange in the dense forests without any habitation for miles and miles. Whiskered terns flying close to a moving launch are a common sight and they aim at fish surfacing on water.

In the recent past 45 varieties of birds have been sighted in the Sajnekhali and Sudhanyakhali areas. Some of these birds like the whimbrel, black-tailed godwit, common sandpiper and common swallow are migratory birds. The local birds include the lesser adjutant stork, redwattled lapwing, ring dove, spotted dove, roseringed parakeet, crow pheasant, green bee-eater, goldenbacked woodpecker, bronzed drongo, blackheaded cuckoo shrike, large cuckoo shrike, whiteeye and blackheaded munia.

Tides occur twice daily in the waters of the Sundarbans. During the monsoon the tidal fluctuation is as much as nine metres. As a result of this tidal action, the creeks are an ideal spawning ground for many kinds of fish. Varieties commonly found are the *bhetki*, hilsa, *parse, bhangan, tengra, pangas* and prawn. Shrimps are also common. Village women are found catching baby prawns with small nets in the rivers and canals near habitation. They earn their living by selling their catch. These prawns are later reared in numerous tanks in the region. The prawn is an important commodity of export these days. The *menomachh* (*Periophthalmus baddarti*) has special air sacs that enable it to stay out of water for sometime. It also climbs trees.

During the winter about 12,000 fishermen catch fish at Jambudwip island not far away from Fraserganj. In the rainy season the Hooghly carries a large volume of flood waters into the sea. The large amounts of biogenic materials contained in these waters produce enormous masses of planktonic algae in the sea which attract crustaceans from the surroundings areas. The planktonic algae and crustaceans provide food to a variety of fish called Bombay duck comprising *herpodon* and *trichinrus* species and *chakul, nuroli, camet, medh* and *phel* varieties. Every year 2000 tonnes of fish are collected, dried and sold through middlemen.

Now, a few words about the extinct animals. The Javan rhinoceros roamed the Sundarbans at one time. This animal is smaller than its counterpart, the Indian rhinoceros. The old official records indicate that the last Javan rhinoceros was killed around 1888. A skeleton of the animal has been preserved in the Calcutta museum. The wild buffalo was found in the eastern part of the Sundarbans till 1885. The animal looks like a domesticated buffalo, but its horn is long. Wild buffaloes are now found in the forests of Assam, Arunachal, Orissa, Maharashtra and Madhya Pradesh. The swamp deer also became extinct by the turn of the present century.

7. Man-eating Behaviour of Tigers

Theories — Taming the man-eater — Electric shock experiment — Conflict between man and tiger — Study of man-eating problem — Relevant factors

Several theories have been advanced to explain the predilection of some Sundarbans tigers for human flesh. One of the theories is based on Dr. Hubert Hendrichs' observation that the number of human killings by tigers 'appears to be positively correlated with the salinity of water.' Incidentally, under the auspices of the World Wildlife Fund, Hendrichs, a German scientist, has studied the man-eater problem in the Bangladesh Sundarbans. Continuous consumption of saline water may have made the tiger irritable by damaging its liver and kidney. Credence is lent to this view by the fact that the man-eating problem peaks in May when the salinity of the water is at its maximum. But there remains a question. As the tiger has been drinking the saline water for ages, its kidney should have developed some kind of ultra-filtration mechanism. No definite answer to this question has yet been found. Eleven tanks have, however, been dug in the reserve to make fresh water available to tigers. It has been observed that tigers visit the tanks regularly.

Another theory, which seems to be more plausible, puts the reason for the tiger's turning man-eater to the disturbance caused by honey-collectors. In April and May the honey collectors roam the forests in small bands in search of combs. Tiger dens in the *hental* clumps are thus disturbed. Taken unawares, a littering tigress attacks a honey-collector in self-defence and becomes a 'circumstantial' man-eater. A defenceless man is an easy prey. Repetition of this kind of incident prompts the tigress to include man amongst her prey or converts her into a 'dedicated' man-eater. The cubs learn from example, which may account for their retaining the habit for life. Hendrichs' report states that only about three per cent of the tigers are man-eaters by preference while another 30 per cent do not go out looking for human victims, but will attack and kill readily, if disturbed. The authorities of the tiger reserve have analysed pug marks of tigers involved in man-killing incidents. The result obtained that four per cent of the tigers are man-eaters, is interestingly, near enough to the earlier finding of Hendrichs.

Recent studies have concluded that in a few micro-regions the existence of a prey-predator gap can also be one of the causes behind tigers turning man-eaters.

Following the killing of an armed forest guard by a tiger in the Chamta block in 1981, a special head-and-nape guard of fibre-glass has been designed to protect people moving in the forests. Interestingly, the head and nape are the targets of the tiger's attack. As the outfit is uncomfortable in summer, it is yet to become acceptable to everyone. This and other measures brought down the annual toll of human lives in the tiger reserve to 29 in 1982 against the average of 45 during the period 1975 to 1982.

In 1983 a new experiment was launched. In the experiment a life-size clay-model of a honey-collector, or fisherman or wood-cutter is placed in a boat near the river-bank or in the forests in an area where a man-eater has been active. (Incidentally, tigers are reported to be non-territorial here as 'scent marking' is not possible. Tides wash away all traces of scent.) Electricity at 300 volts is passed through a galvanized wire round the neck of the model from an energizer connected to a 12-volt car battery. A fuse recorder is put in the circuit. The problem lies in deceiving the tiger into attacking the model. Clothes recently worn by a honey-collector, fisherman or wood-cutter are used on the model as an olfactory lure. In June 1983, for the first time, a man-eater suffered an electric shock when it jumped on a fisherman's model in a country boat at Sudhanyakhali. The next attack was on a wood-cutter's model at Netidhopani in February, 1984 by a man-eater who had accounted for 24 deaths. When a team of forest officials visited the site the following morning, the man-eater called out in pain and scampered into the dense forest. It is claimed that the electric shock does not seriously hurt the 'shocked' man-eater. The case of the 'shocking' of the 'Kali Char man-eater' is fascinating. The man-eater residing at Kali Char at the confluence of Pichakhali and Dattar Chhera rivers had killed over 60 men. Every year the first tiger-kill was reported from Kali Char. In early January, 1985 the man-eater attacked again. Fortunately, he picked a clay-model, not a living man. As the model was not energized, it was shattered to bits. It is strange that the man-killing at Kali Char stopped after the incident. Perhaps the tiger didn't like the taste of 'human flesh' and turned to normal prey thereafter! So far nine man-eaters have been 'shocked' in the Sundarbans.

In the latest experiment the mask of a human face is worn by a honey-collector on the back of his head to confuse the tiger who normally attacks from behind a man. The annual average of human killings over the past five years has been 31.

The conflict between man and tiger in the Sundarbans originates mainly in the socio-economic condition of the local people. In this region scheduled castes and tribes account for 42 per cent of the population, against 26 per cent in the state of West Bengal. The per capita income is estimated to be less than half that of the state average. In the struggle for survival, every year some 4,000 fishermen, 500 honey collectors and 500 wood-cutters of the region enter the forests, braving the lurking sharks and crocodiles in the rivers and the diabolical tigers on land. Collection of honey, wood and fish is allowed only in the buffer area of the tiger reserve, no disturbance being allowed in the core area. These activities are controlled with permits by the project authorities, who also provide armed escort to the permit-holders.

A notable feature of the Sundarbans is that there are no enclaves of cultivation anywhere. Depredation by tigers in the adjoining villages which are separated from the forests by rivers and channels, has not generally posed a serious problem. In 1984, though, a tiger swam across the river Punjali from the forest block of Jhilla and killed 20 head of cattle. An electric fence more than a kilometre long, was erected along the river bank. This succeeded in deterring the tiger, though when the fence was removed, he reappeared. Subsequently, the fence was repositioned but not charged electrically. The tiger has not visited the village again!

A recent study has made a number of recommendations for reducing areas of conflict between man and tiger. Some of the recommendations are as follows : (a) Permits for *hental* leaves should not be granted. This recommendation has been enforced. (b) In order to identify the man-eaters, pug marks of the tigers involved in incidents of man-killing should be collected and analysed. This recommendation is being implemented. (c) The 'tiger guard' made of fibre-glass should be used by person working in the forests. As mentioned elsewhere, the outfit is yet to become popular. (d) In order to fill the prey-predator gap, pigs of wild strains should be released in the concerned pockets. A piggery is now being maintained at Pakhiralay. (e) In order to win the confidence of villagers who have been harassed by straying tigers, the method of immobilizing tigers and transferring them elsewhere should be used. The method was used in some incidents of straying in the past, but in recent years, errant tigers have been trapped and released in the core area. The method of chemical immobilization involves nursing the animal for sometime in the zoo. (f) More fresh water tanks should be dug inside the forests. This is being implemented. (g) The experiment of subjecting man-eaters to electric shocks should be continued. This has been accepted.

River and swamps : the ubiquitous features of the Sundarbans

Measures against man-eaters : energized clay model of a wood-cutter.
Photograph by Pranabesh Sanyal

Deer by a freshwater pond at Sajnekhali.
Photograph by Pranabesh Sanyal

A 'Royal' Bengal tiger at the Sundarbans.
Photograph by Pranabesh Sanyal

An estuarine crocodile on a mud-flat at Sundarkhali

Open-bill storks at the Sajnekhali Bird Sanctuary

Sundar trees near the Haldi watch-tower

The watch-tower at Netidhopani

8. Stories of Straying Tiger

An unwelcome guest at Jharkhali — The long march of a tigress — 'Sundari' of the Calcutta zoo

An Unwelcome Guest at Jharkhali

In the early morning of August 2,1974 a tiger killed a woman in the Jharkhali village, 70 kms. from Calcutta. The tiger did not eat the woman's body. At the time of the incident the woman was seated near a goat-shed. The circumstances suggested that the woman was an accidental victim.

The site of Jharkhali was part of the reserve forests until 1955 when it was assigned for the settlement of refugees. There are still forests on the south of the village. The tiger was sighted frequently in the next few days and the villagers were alarmed. The narrow belt of mangroves along the Matla and the large marsh located in the centre of the village, provided shelter to the animal. The tiger also killed some dogs, cattle and chicken. In most cases the animal was unable to feed on its kill because of the alarm raised by the villagers. It didn't attempt to defend its kill, nor did it return to its kill.

The villagers demanded that the tiger be killed immediately. However, evidence indicated that the animal was not a man-eater. So the West Bengal Forest Department decided to capture the animal by darting and translocate it. The method of tranquilizing tigers had been used in Africa and North America, but had never been used in India. At that time Seidensticker, a tiger expert was participating in a workshop on tiger-ecology in the Nepal terai. At the request of the West Bengal Government, Seidensticker flew down from Nepal.

At Jharkhali, the reclaimed area is maintained with a system of embankments which also provide pathways for the local people. The embankments had been breached in many places on account of the heavy monsoon. Consequently some areas were innundated during the high tides. The tiger was soon located at a hummock in the marsh.

In the next nine days the team of forest officials including Seidensticker tied three bullocks, one at a time, on the hummock as bait. The tiger killed and ate the first two baits. It also killed two dogs who came to feed on the carcass. A boat with a secure hide was placed in the marsh about 15 metres from the bait. A five metre-high platform was also erected for fixing a spotlight.

The tiger didn't visit the third bait until the mid-night of August 26. A few minutes after it had killed the bullock, it was darted. It was found immobilized ten metres away on a mud–flat. The animal was found to be a young male without any noticeable physical defects. So, there was no good reason for keeping him in captivity. It was decided that he would be released in the core area of the tiger reserve.

The animal was put into a cage which was transferred to a launch. In about two hours, the effects of the drug wore off and the animal started pushing the bars of the cage. At the site of release, the cage-door was opened, being operated by a rope from the launch. The tiger walked out and soon disappeared into the dense forests.

The local forest officials maintained surveillance on the tiger during the next few days. The tiger was seen returning to the cage on August 28 afternoon, and leaving it the next morning. On receipt of this information the team led by Seidensticker rushed to the place and was shocked to find the dead body of the tiger about 20 metres from the cage. There were many pug marks of tiger all round, some of them being larger than those of the dead tiger. The team concluded that the tiger had died of injuries resulting from an encounter with a larger tiger. That was the tragic end of the first experiment of tranquillizing a straying tiger in the Sundarbans.

The Long March of a Tigress

On November 10, 1985 the inhabitants of the Punjali village near Gosaba woke up at dawn to discover that a tiger had carried off a sheep. The tell-tale blood-trail and pug marks were proof of the attack, but remnants of the kill could not be recovered. Punjali is 8 kms. from the Jhilla block of the Sundarbans forest, with two rivers, Korankhali and Punjali, in between. The village of some 800 households is fairly prosperous by local standards. Apart from a few thorn bushes around homesteads, it has little tree–cover. At that time the fields were aglow with golden paddy which was to be harvested soon.

The assistant field director of the Sundarbans Tiger Reserve arrived at the village the next day. The pug marks told him that the big cat was a tigress. He first sighted the animal on the 12th. It was believed that the tigress had strayed from the Jhilla block of the Sundarbans and crossed the two rivers on her way (See map). She took cover in the paddy fields and her dreadful presence was betrayed by the sudden swaying of rice plants.

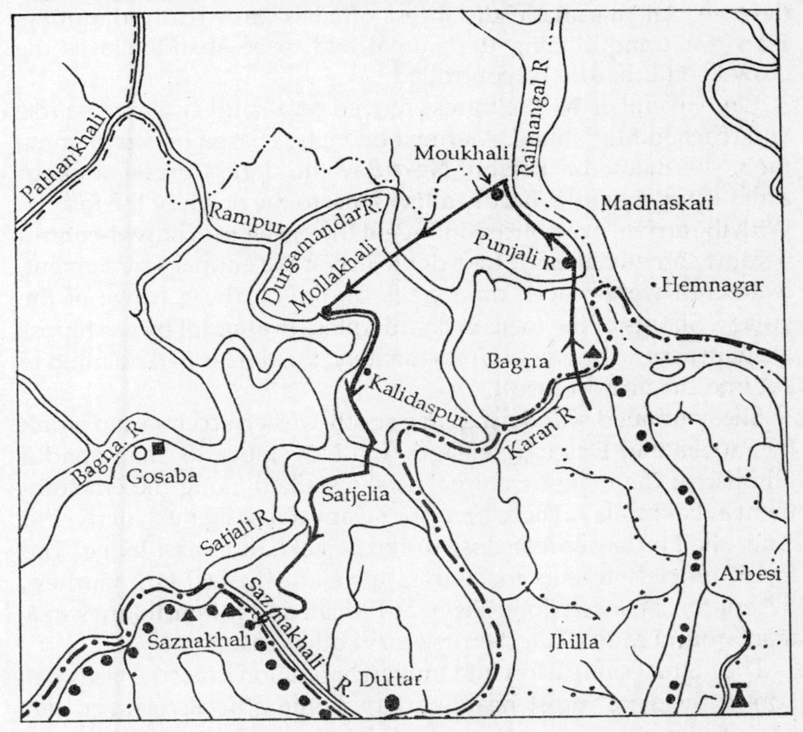

Forest Boundary

Bird Sanctuary

The forest officials were besieged by angry and excited villagers and could not implement their plan of driving the animal back to the Jhilla forest. On the 13th, the field director armed with tranquillizing equipment and a trap arrived at Punjali. The trap was laid with the bait of a live goat. In order to prevent the animal from entering other villages, the forest officials and villagers organized patrols on the northern outskirts of the village. The patrols used crackers and high–power torches. But the tigress sneaked through the cordon at night and moved to Moukhali, 17 kms. inside human habitation. The residents had not come across any tiger in the village in their living memory.

Very soon a large crowd gathered to have a glimpse of the unwelcome guest. Tension ran high and some men pelted stones at the animal. The unnerved tigress could have mauled anyone in self-defence. All this made the forest officials' task really daunting. Plans for tranquillizing the animal had to be abandoned as the crowd could hardly be controlled.

On November 16, the tigress moved on a south-western course and reached Mollakhali. Alarmed on being chased by a screaming mob, she mauled a person. Next day the tigress prowled about Kalidaspur, a hamlet hardly a few kms. to the north of the forests. With the arrival of a police contingent the problem of crowd-control became somewhat easy. In order to arrest the animal's movement, watercraft were placed on a creek on the northern fringe of the village. Shouts of the men on board foiled an attempt by the tigress to swim across the creek. Unfortunately, the tigress had no mind to get into the nearest forest.

She continued wandering on a south-western course and slunk by Satjelia and Dayapur villages. On November 18, many had a glimpse of the majestic animal as she stalked along the embankment at Pakhiralay. Thereafter, she swam across Pichekhali river to enter the Pirkhali forest close to the Sajnekhali tourist lodge. The forest officials trailed the tigress all along her 60-km. journey. Except for the loss of one sheep at Punjali and minor injuries to a man around Moukhali, there were no other casualties.

The animal's intrusion into human habitation caused consternation in the state forest headquarters. Some officials claimed the tigress's long journey proved that the species was wandering in nature. In support of the claim they emphasised that there was no dearth of prey like pigs and deer in the Jhilla forest from where the tigress had strayed.

The tigress again came the notice of forest officials on March 19, 1986, at Lahiripur. Some women were fishing near the Duttar forest checkpost, when the tigress suddenly appeared and mauled one of them. The woman subsquently died in the Gosaba hospital. The tigress next appeared at Sudhanyakhali on April 3, and carried off a fisherman. His body could not be recovered. The tigress was identified by her pug marks.

On several occasions from the beginning of March 1986, the tigress killed cattle, goats and pigs at Jamespur and Dayapur villages on the bank of Pichekhali river opposite Sajnekhali. On her way she traversed the forest around the Sajnekhali tourist lodge and visitors had the thrill of seeing her on a number of occasions. The villagers were panic-stricken and thier rage against the forest officials mounted.

Towards the end of March an attempt by the forest officials to trap the animal with the bait of a live goat at Jamespur failed. The villagers were at the end of their tether and sent a deputation to the forest department in Calcutta. The department directed that the animal be trapped and released deep inside the forests. In view of the recent death of a tiger after being tranquillized at Hiranmoypur, this process was not recommended.

On April 20, 1986 a trap with a wild pig as the bait was laid in the forest about 200 metres to the north-east of the Sajnekhali tourist lodge. Forest officials stationed in a nearby launch kept vigil. Simultaneously, an electric fence was erected along the bank of Pichekali to prevent the animal from entering the villages. The tigress was perhaps hungry and walked into the trap in the early hours of April 22. The forest headquarters in Calcutta instructed the local officials to release the tigress around Haldi block in the core area of the tiger reserve.

The tigress was transferred to a lighter trap to facilitate handling. In the morning of April 23, the launch *Manorama* with her Royal Bengal passenger cruised towards Haldi. Meanwhile, the animal had finished the bait and also drunk some water. No veterinary aid was readily available but the tigress looked cheerful. She enjoyed another meal of wild pig that was offered.

The *Manorama* reached Narayantala in the vicinity of the Haldi watch tower at around 4 pm. The trap was placed on the mud–flat under a tall *keora* tree from the branches of which men could operate the trap-door. The forest officials were ready for release of the animal, but under instruction from the headquarters, they awaited arrival of the field director, who had rushed from Calcutta.

With the arrival of the flood-tide, the water started rising. The mud-flat was inundated. The tigress was in a pitiable condition, and the water was almost touching her neck when the field director's launch finally arrived. Dusk had set in and visibility was poor.

Two forest officials, Maity and Tanti, well-known for their courage and dedication, volunteered to operate the trap-door. They swam through the swirling waters, braving the menacing crocodiles and a pair of tigers said to be on the prowl in the area. As they lifted the trap-door from their perch on the tree, the drowning tigress jumped out and bounded beyond the edge of the water. Before slinking away into the forest, she had a lingering look at her benefactors on board. She was hardly aware that some of her saviours were on the tree.

For many weeks the tigress's movements were monitored by the officials of the Haldi watch tower and her pug marks were detected at regular intervals on the banks of the fresh tank adjoining the tower. It was a happy end to a traumatic chapter of her life.

'Sundari' of the Calcutta Zoo

In the incidents of straying in the Sundarbans two tigers and one tigress have so far been tranquillized and transferred to the zoo in Calcutta by the forest officials. These tigers have since been living happily in the zoo. In the last of the three incidents the tigress was immobilized in the Uttardanga village near Gosaba on January 4, 1982. The animal had entered a cowshed in the village. Around eight in the evening the forest officials reached the disembarkation point in their launch. After walking about a kilometre on the embankments, they arrived at the village. The villagers had already shut the door of the cowshed and secured it with strong fishing nets to prevent the animal's escape. In the presence of about a thousand people, the darting of the tigress began. As many as three darts were used for immobilizing the animal. She was tied to bamboo posts and carried to the embarkation point. She regained her senses only after three hours. Now, she is known as 'Sundari' to the visitors of the Calcutta zoo and has recently given birth to two cubs.

9. Tiger Census

Tiger census in the Sundarbans

It is well known that Jim Corbett always used pug marks for tracking down man-eating tigers. Sankhala has mentioned that the tiger is a soft padded animal who prefers to walk on soft soil and clear paths to avoid injury to the paws by thorns and sharp stones. Census of tiger is based on impressions of paws or pug marks. The pug marks of one tiger are distinguishable from those of another. The first enumeration of tigers based on pug marks was perhaps conducted by W. J. Nicholson of the Imperial Forest Service in Palamau in Bihar in 1934 and 32 animals were found in an area of 299 sq. kms.

In order to compare pug marks, their tracings are taken on glass-plates and later transferred on paper along with other data. When these tracings and data are subjected to analysis, it is possible to identify precisely an individual tiger. The equipment for the purpose, evolved by Mr. S. R. Choudhary, is called the 'Tiger Tracer'. It is an extremely simple device which makes accurate tracing of pug marks possible and consists of a colourless rectangular glass-plate of size 20 cms. x 25 cms. x 3 mms. It has four holes near the corners and each of this carries a screw held in place by a nut and washer, each screw having an extra fly nut. A few rubber bands that can wrap round the plate, a free flowing pen (such as a fibre micro-tip pen), some papers and a metre-long steel tape complete the equipment.

Like all other quadrupeds, a tiger in its normal walk, places the hind paw of one side at the same place where the front paw of the same side earlier rested. This results in partial or complete superimposition on the fore pug marks of the hind pug marks. When the animal moves faster the hind foot over-shoots the front paw and in such a situation it is possible to have the marks of both front and hind paws. However, in any situation the hind pug marks can be seen intact and so the hind pug marks are used in census-work.

There is yet another reason for using the hind pug marks for enumeration. The front limbs of a predator are its main weapon for attacking its prey. The forearms of the front limb, including the toes, claws and pad, are used in gripping and pulling down the prey and are accordingly massively built in both the sexes. The forepaw is therefore substantially larger than the rear one in either sex and a tracing of the pug mark, if enclosed in a linear frame, tends to be squarish in shape. However, if a tracing of the hind pug mark is taken, that of the male fits a squarish frame, whereas that of the female a rectangular frame. Thus, the hind pug marks enable the enumerator to determine the sex of an individual tiger.

As the paw is a supple organ, it adjusts itself to the ground available for resting it. It can appear bigger on a slushy and looks smaller on a hard sub-surface with a thin cover of dust or sand. A small article of gravel or a small hole can cause the pug mark to twist on one side. As far as possible, a perfectly normal rear pug mark should be selected for tracing.

It is interesting to note that only four toes form an impression along with the pad in the rear pug mark, the fifth toe being well above the pad, does not touch the ground in a normal walk. The retractile claws also do not come out at the time of walk and therefore, leave no imprint, except when the tiger is walking on slushy ground.

The glass-plate of the tiger tracer is placed above the pug mark. The outline of the pug mark should be traced taking care to avoid parallax error by moving the eyes vertically above the segment under tracing.

The tiger is an individualistic animal and associations between adults occur only during courtship. Rearing the cubs is the responsibility of the mother and therefore the mother-cub association is quite usual. Confusion may arise while identifying the pug marks of a tiger-cub below 6 months of age and those of a leopard. Here, one should remember that a tiger cub never moves alone and always trails his or her mother. Small pug marks unaccompanied by bigger pug marks can, therefore, be presumed to be that of a leopard.

In 1984 the tiger-census in the Sundarbans continued for seven days. During this period teams of enumerators entered the creeks when the tide was high, and collected the pug marks from the adjoining mud-flats. In the entire Sundarbans including the areas outside the tiger reserve 287 tigers — 150 males, 120 females and the rest cubs — were found. The work of tiger-census began again on November 30, 1988. Unfortunately, an unprecedented cyclone, which occurred on that day, played havoc with the enumerators. A departmental launch with two officers on board was drowned and two other officials died. The census work was suspended and completed only recently.

10. Other Conservation Experiments

Olive Ridley turtle —Fishing cat — Estuarine crocodile — Bhagabatpur rearing station and breeding farm — Horseshoe crab — Batagur terrapin— Hawksbill turtle

The Olive Ridley turtle (*Lepidochelys olivacea*), an endangered species, visits the beaches in the Sundarbans in December to lay eggs in the sand. These eggs fall prey to men, water monitors and wild pigs. Hatchlings on their journey back to the sea are devoured by sea gulls, whiskered terns, brahminy kites and other predators. As a conservation measure, eggs are now collected from Mechua and the Ridley islands on the seaface and hatched in artificial nests at Sajnekhali and Bhagabatpur. When the young ones are two months old, they are released in the creeks near the Bay. About 2,000 turtles have been released since 1983. Incidentally, in the past, the animal was hunted indiscrimately for its meat. It is reported that the turtle swims as long as 3,500 kms. to find nesting ground.

A few young of the fishing cat were found abandoned by their mothers at Bagna, Jhilla and Sudhanyakhali. They were rescued and experiments on their food habits and development are now being conducted by the project authorities at Sajnekhali.

The estuarine crocodile proliferated in the Sundarbans in the past. But the animal was mercilessly killed for its skin and now it is a threatened species. In view of the key role of the animal in maintaining the mangrove ecology, special conservation efforts are needed for its preservation. In May-June the female crocodile makes nests on river-banks with *hental* leaves. Heat from the fermenting leaves helps incubate the eggs. The female generally remains in the vicinity of the nest to guard the eggs. The nests are about 30 metres from the edge of the water. The female scoops out a basin near the nest. When it is inundated during the high tide, she splashes water with her tail over the nest so that the eggs are embedded in the earth. During May and June, forest officials go out in search of the nests. After a nest is located, each egg is collected carefully by hand in a wooden box, after marking its top and orientation with the help of a compass. The eggs are then brought to Bhagabatpur where they are hatched in artificial nests, each egg being placed with its original orientation. The period of incubation lasts 70 to 90 days and as soon as the embryos start croaking, the nest is broken and the hatchlings are taken out. When the young ones are two years old, they are released in the rivers of the Sundarbans, 146 crocodiles having been released since 1979.

A crocodile breeding farm has also been set up at Bhagabatpur. For this purpose three male and four female crocodiles have been kept in the three tanks there. Thirty-eight hatchlings have so far been obtained from the farm-bred eggs. Incidentally, some countries like Thailand, Japan, Indonesia and Singapore also have crocodile breeding farms.

The horseshoe crab has recently been taken up for study. The *Batagur baska*, an endangered terrapin is also being studied. This species was hunted mercilessly in the mid-nineteenth century for its fat which was used in the manufacture of soap in Calcutta. Another endangered species, the hawksbill turtle has, recently been found in the Sundarbans.

11. Economy

The Economic importance of Sundarbans — The Sundarbans development

The Economic Importance of the Sundarbans :

It is well-known that the forests of the Sundarbans provide a buffer to inhabited areas against the ravages of cyclones originating in the Bay of Bengal. The forests also play an important role in controlling atmospheric pollution. In addition, the mangroves of the region perform three useful functions. Some plants like the *khalsi* (*Aegiceras carniculatum*) and *harguza* (*Acanthus ilicifolius*) have nectar-bearing flowers. Rock-bees from the Himalayas visit the forests every year and make bee-hives with nectar from these plants. In April and May, honey-collectors collect, on an average, 500 quintals of honey and 30 quintals of wax. As the density of honey depends on the number of salt-excretory glands, the khalsi, having 19 glands per sq. mm., gives the best honey. As rock-bees are migratory, the experiment of setting up apiaries has failed.

Secondly, scientists have found that the estuary is more productive than the sea or fresh water river. The tides provide energy for rapid cycling of nutrients, resulting in the rapid development of an organism to maturity. While preparing the management plan for the Bangladesh Sundarbans, Dr W. W. Odum and Dr E. Heald have studied the relationship between the terrestrial primary production of the mangroves and the aquatic secondary production. They have found that the mangroves produce a large quantity of organic matters which sink deep into water. Some varieties of fish like the *pangas* and *kanmagur* directly eat fruits of the *keora* and *baen*. As the organic matters decompose, they are turned into detritus particles covered with bacteria, protozoa and fungi. During the decomposition plenty of carbon-dioxide is released. As a result, photoplanktons multiply and zooplanktons, in turn, proliferate with the help of oxygen released by photoplanktons. In winter the bioluminiscene of noctaluca floating on water in the Sundarbans estuaries attract the visitor's attention. Shrimps, crabs and *parse* feed on the zooplanktons while many aquatic and some terrestrial animals feed on them. Many varieties of sea-fish spend a part of their lifetime in the estuaries. Estuaries are the ideal spawning ground for many varieties of fish. Experts believe that the excreta of crocodiles provide considerable amount of nutrients to aquatic animals. The local fishermen catch about 120 varieties of fish in the region. It has also been established that coastal fishing depends on the mangroves to a large extent.

Thirdly, the Sundarbans forests are a source of fuel and small wood to the local people, the annual outturn being 100,000 quintals. A special feature of the small wood of mangroves is its anti-termite property.

The Sundarbans Development :

The inhabited area of the Sundarbans is as large as 6,630 sq. kms., with a population of 2.6 million. The area has numerous rivers, canals and creeks. The water, being brackish, is unsuitable for irrigation and the land being low has poor drainage. A recent study reports that the daily income of the local people is as low as Rs. 5 to Rs.12 and that, too, depends on the tides, floods and erratic monsoons. Almost three-quarters of the population do not have regular jobs and live on part-time occupations like fishing and selling fire-wood and honey. The literacy-rate is hardly 25 per cent. Canning, Namkhana and Patharpratima are the only small towns that cater for the population who make weekly or monthly visits by boat to buy provisions, sell their produce or stay overnight to watch a *jatra* (open-air theatre).

Agriculture is the staple occupation of the local people. In the absence of irrigation, land is mono-cropped with summer rice. A World Bank-assisted project with an outlay of Rs. 320 million has been launched for speeding up the region's development. The thrust of the project is on agriculture, aiming at two or three crops a year, including cash crops like water-melon, chilli, sunflower, radish and sugarbeet. The other areas of thrust are communications, irrigation, drainage, pisciculture, social foresty, market development and education. Since the region has no major industry and power is not available, activities relating to cottage and small industry, poultry and animal husbandry are also being encouraged.

In view of the undeveloped communication facilities in the region, the project has not progressed satisfactorily. Many new roads, bridge, jetties, sluices and canals are under construction. A new jetty has been commissioned at Sonakhali and a fish farm at Jharkhali. The social foresty activities include planting mangroves on river-banks, planting trees on road-sides and distributing seedlings of fuel-wood and fruit species to small farmers, free of cost.

Under the auspices of the Sundarbans Development Board, the outline of a Rs. 5,000 million project has recently been prepared. In this project irrigation, flood-control, pisciculture, social foresty, the power and paper industries have been accorded priority. As the earlier proposal for a floating hotel in the forests has now been shelved, need for rapid transportation of the tourists to the Sundarbans has been emphasized; use of helicopter, hovercraft and hydrofoil has been recommended. Recently a plant for the production of industrial alcohol has been opened at Nimpith. The plant uses sugarbeet, which has been introduced in the region not long ago.

12. Deities of the Sundarbans

Ban Bibi — Dakshin Ray — Mobrah Ghazi — Fakirs

The guardian deity of the forests of the Sundarbans is Ban Bibi (lady of the forests). Ban Bibi, having great love and affection for her devotees, is ever eager to protect them. That is why clay-modellers always portray the deity as a pretty and graceful lady. In some places the deity is mounted on a tiger or a hen and dressed like a Muslim woman with ghagra-choli on her body, plaited hair and brocade-cap on her head, having a child in her arms. In other places the deity is in Hindu dress. There is no especially auspicious day for worshipping the goddess. Some people worship her before entering the forests while others do so after returning safely from the forests.

The legend of Ban Bibi reads like this : A merchant of Kalinga city set out for the Sundarbans in a boat to collect honey and wax. His nephew, Dukhe, the only son of his widowed mother, accompanied him. While the boy left for the dreaded forests his mother was in tears and prayed to Ban Bibi for his protection. The merchant went deep into the forests and landed there after offering prayers to Dakshin Ray, god of the tiger, while Dukhe stayed in the boat. The merchant along with his associates could not locate any beehive as Dakshin Ray played a trick with him. The despondent merchant returned to the boat and fell asleep. Dakshin Ray appeared in his dream and proposed to give him enough honey and wax if Dukhe was sacrificed to him. After a little hesitation the merchant agreed to the proposal and his boat was filled with honey and wax. On the way home Dukhe was thrown overboard and with great difficulty he managed to reach the river-bank. Dakshin Ray appeared in the garb of a tiger and was about to devour him when Dukhe started praying to Ban Bibi who appeared on the scene in no time and took Dukhe in her arms. Dakshin Ray fled and at the command of Ban Bibi, her brother, Jangali, drove him out of the forests. Dakshin Ray approached Ghazi Saheb and begged him to intercede on his behalf. On the advice of Ghazi Saheb, Ban Bibi pardoned Dakshin Ray.

In the Sundarbans Dakshin Ray is the god of the tiger. He is generally worshipped below a banyan, peepul or *nim* tree, the image being a mound of earth or a block of stone painted with vermilion or a queer-looking head. Wood-cutters, boatmen, honey-collectors and other communities who depend on the forests for their livelihood worship this god, the most auspicious day being on *Makara Sankranti* (occurring in the middle of January).

The legends of Dakshin Ray are found in the *Raymangal* which was composed by some poets in the middle ages. The book conceives Dakshin Ray as a warrior seated on a tiger with a bow and arrow in his hands. However, this image is not used by the local people. The *Raymangal* records as follows : Puspa Datta, a merchant of Bardaha asked Ratai Baulya to make some commercial boats. Ratai and his six brothers went deep into the forests and collected a lot of timber for the boats. As they were about to return, they came across a large tree and immediately felled it. The tree was the abode of Dakshin Ray who got angry and got all of them, except Ratai and his son, killed by tigers. Ratai was about to commit suicide when a heavenly voice asked him to sacrifice his son to Dakshin Ray in order to resuscitate his brothers. Ratai obeyed the advice and his six brothers regained their life. Ratai and his brothers returned and narrated the supernatural powers of Dakshin Ray to the local people.

Puspa Datta offered a bag of gold coins to the craftsman who could make the boats. At the command of the master of heaven, Mahadeva, the two craftsmen Hanuman and Biswakarma, decended to the earth in the garb of human beings and made seven boats in the next seven days. The best boat was named the *Mudhukar*. Puspa Datta wanted to set out on an arduous journey across the sea in search of his missing father. Sushila, Puspa Datta's mother offered prayers to Dakshin Ray and entreated him to save her son from any danger. She advised Puspa to pray to Dakshin Ray whenever there was any danger.

Puspa Datta began the journey aboard the *Madhukar*. On the way he worshipped Siva at Barasat and reached Khania where he offered prayers to Dakshin Ray. There he came across an altar of Ghazi Saheb. On enquiry, the boatmen explained that once when there was a confrontation between Daksin Ray and Ghazi Saheb, one could not defeat the other. Realizing that the continued fighting meant destruction of the whole earth, god appeared in the garb of a saint and mediated between them. In accordance with the agreement, Dakshin Ray was to be the master of the southern districts and Ghazi Saheb the master of the Hijli region. It was also agreed that every devotee would pay the same respects to Dakshin Ray and Ghazi Saheb.

Then, Puspa Datta arrived at Ganga Sagar and heard the story of Bhagirath bringing the Ganga to the earth. He sailed along the coasts and at Rajdaha came across a wonderful sight —- a beautiful palace on the sea. His companions, however, failed to view the sight. Finally, Puspa Datta arrived at Turanga where he narrated to the king his experience of the wonderful sight. On hearing this unusual story, the king scolded him. When Puspa Datta failed to show him the wonderful sight, he was imprisoned and ordered to be beheaded.

Puspa Datta prayed to Dakshin Ray. Next day when he was being taken to be executed, Dakshin Ray along with a few tigers appeared there. The army of tigers captured the city and the king lost his life to Dakshin Ray. The queen arrived at the battlefield and heard a heavenly voice — her husband would be resuscitated, if her daughter was married to Puspa Datta and she worshipped Dakshin Ray. The queen agreed and the king and his soldiers regained their life. Puspa's father had been imprisoned by the king. Puspa got his father released and married the princess, Ratnabati. Ultimately, Puspa Datta along with his father and newly-wedded wife, returned to his country. Puspa's father, Deva Datta became a devotee of Dakshin Ray. Thus, the greatness of Dakshin Ray spread far and wide.

The adoration of *pirs* or Muslim saints is common among the people, both Hindus and Muslims, of the Sundarbans. The most famous *pir* of the region is Ghazi Saheb whose supernatural powers are described in the following legend. The area in the Maidanmal *pargana*, south of Tolly's Nullah and containing Baruipur, was earlier a dense forest with wild animals. A fakir named Mobrah Ghazi took up his residence in the forest. He overawed the wild animals and always moved about the forests on a tiger. Once the local zamindar was unable to pay his renenue. The emperor ordered him to be arrested and brought to Delhi. The zamindar's mother sought the fakir's assistance in getting her son released. The emperor had a dream in which Mobrah Ghazi asked him to release the zamindar saying that he was the proprietor of the Maidanmal jungle and that the revenue due from the zamindar would be paid from his treasures buried in the forest. The next morning when the emperor ascended his throne, he found himself surrounded by wild animals. He at once ordered the release of the zamindar and sent him back to Maidanmal instructing him to locate the treasures. On reaching home, the zamindar informed his mother of all that had happened. His mother went to Mobrah Ghazi who pointed out the place where the treasures lay and disappeared mysteriously. The treasures were dug up and the revenue due to the emperor remitted. The zamindar wished to erect a mosque in honour of Mobrah Ghazi who observed that he preferred living in the forests and receiving offerings from all who came to cut wood.Thereupon, the zamindar ordered that evey village should have an altar dedicated to Mobrah Ghazi, and warned his tenants that if they failed to make offerings before entering the forests, they would be devoured by tigers. Mobrah Ghazi's altars are common in villages in the vicinity of the forests.

A number of fakirs, who call themselves descendants of Mobrah Ghazi, earn their livelihood by the offerings made by wood-cutters, honey-collectors and boatmen for their services of providing protection against tiger-attacks. The custom is for the fakir to accompany the persons to the spot where they have to work, and make a clearing repeating charms and incantations. The fakir builds seven small huts with stakes and leaves within the clearing. The first five huts are dedicated to the gods and goddesses of the Hindu pantheon. The sixth hut is reserved for Ghazi Saheb and his brother Kalu, and the seventh for his son, Chawal Pir, and his nephew, Ram Ghazi. The offerings consist of rice, plantains, coconuts, sugar, sweets etc.; small earthen lamps are also lit. The fakir, after having a bath, wears a new *dhoti* with his hands, arms and forehead smeared with vermilion. After offering prayers before all the deities, the fakir proceeds to ascertain whether a tiger is present in the locality or not. If a tiger is in the vicinity, the ground under his feet feels warm and his head is involuntarily turned in the direction of the tiger. The fakir 'drives' away the tiger by repeating incantations. In order to close the eyes of the tiger, he repeats the incantations, 'Dust! Dust! The finest dust be on thy eyes, Oh tiger and tigress'.

O'Malley's gazetteer of 24-Parganas has recorded : The wood-cutters have great faith in fakirs. The fakir's charms and exorcisms give them courage to enter the forests and work there. However, that his charms and incantations have little effect has been proved, for it often happens that the fakir himself is carried off by the tiger.

13. Remains of Old Civilization

Chandraketugarh — Netidhopani

The recent exploration and excavation at Chandraketugarh near Berachampa in the North 24-Parganas district has shed light on an invaluable chapter of Bengal's history. Some historians believe that the city of Gange, the royal residence of the Gangaridai mentioned by Ptolemy, may have been at Deganga. Interestingly enough, Chandraketugarh is in the Deganga block. The ruins of a walled city, 6 sq. kms. in area, along with many significant relics have been discovered. Of the relics, which date to a period between the Maurya (4th to 3rd century B.C.) and Gupta (4th to 6th century A.D.) ages, mention may be made of some silver punch-marked coins, copper coins, polished pottery, terracota figurines, and a rare Gupta gold coin. The unique terracota *Surya* chariot may also be mentioned. The earliest Budhist image, so far discovered in Bengal, has been found at *Khana-Mihirer Dhipi* near Chandraketugarh.

The finds at Chandraketugarh, situated on the bank of the dried-up river, Bidyadhari, once a large branch of the Bhagirathi, point to the existence of a flourishing international port rivalling the ancient port of Tamralipta in the Midnapore district. The importance of the Bidyadhari as a maritime route flanked by cities, temples and monastries is supported by the explorations downstream at Khas-Balanda, Dhara and Bhangor. The ruins of a Gupta stone temple, later converted into a mosque, have been discovered at Haroa, 10 kms. from Berachampa.

Remains of a couple of old temples have been found in Sagar Island. Two large old tanks, Raydighi and Kankandighi, were discovered to the west of Hatiagarh. In the adjoining Lot No. 116 is a famous temple. Nothing definite is known about the history of the temple. In Lot No. 127 there is a large masonry structure called "The temple of Birinchi'. An idol of Siva and remains of a temple have been excavated in Lot No. 101 west of Kalindi, off Kaliganj. Henckelganj (popularly called Hingalganj), situated to the north of this place, is a settlement that was set up and named after the first Judge-Magistrate of Jessore.

At Netidhopani remains of a temple and a row of houses have been discovered. Rows of exotic *Bokul* (*Minusops hexadra*) trees found there are said to have once beautified a road. The place is believed to have been a halting station of the legendary merchant, Chand Saudagar, on his way to Sagar Island. The snake goddess Manasa was first worshipped at Netidhopani. The place also figures in Rennel's map dating to the eighteenth century.

The large tanks, masonry structures and high embankments, found at many places in the forests, are cited as indications of a former prosperity of the area. It is well known that in the past dacoits and pirates had their retreats in the forests from where they emerged to raid the country and waters around. Moreover, during the first half of the nineteenth century the government had centres for manufacturing salt at various places in the forests. In the opinion of *The Calcutta Review*, July 1889, "these two facts go some way to explain the existence of many masonry remains without resorting to the theory of a former widespread civilization".

14. Snippets from Old Pages

The local woodcutters refer to the tiger as the *baraminya* (senior headman) and the crocodile as *chhotominya* (junior headman).

In the Sundarbans when the river flows through habitation, stakes are driven into the bed of the river at the places where the inhabitants bathe and draw water for domestic purposes. But even this precaution is not always sufficient to ward off attacks by the fierce crocodiles. The crocodile, being an amphibious animal, finds no difficulty, when pinched by hunger, in turning the flank of the stakes, and taking up its post within the enclosure, where it silently awaits its prey. A friend of the Morrieson brothers had witnessed the following shocking incident. A young Hindu girl, about fourteen years old, came to get a pitcher of water, and had hardly put her feet into the water, when a crocodile, who had been lying in wait inside the enclosure, rushed at the poor girl, seized her in its formidable jaws, scrambled up the banks of the river, holding the shrieking, struggling girl well up in the air by the middle of her body, and plunged into the water outside the stakes. With smothered scream, a ripple upon the water and a few bubbles the frightful scene was closed.

In a district-town a group of convicts in iron were being inspected by the magistrate prior to their being sent off to a distant jail. The men along with their guards, numbering about 50, were drawn up in a line on the raised embankment of the river. When the inspection was in progress, a crocodile rushed up the bank, seized an ironed convict by the legs, and in a moment plunged into the river and disappeared.

(*The Calcutta Review*, March 1859)

The tigers stalk deer and pigs and will swim broad streams in their search for prey. Wild buffaloes are found in the eastern parts and rhinoceros in the depth of the forest near the sea coast.

The chief local divinities are Chawal Pir (boy saint) and Ban Bibi (Lady of the forest). Both seem to be of Mohammedan origin, but they are worshipped by both Mohammedans and Hindus and their shrines are often marked by nothing more than a bamboo topped by a flag.

(*The Calcutta Review*, July 1889)

Ghazi Saheb and his brother Kalu, are venerated both by Mohammedans and Hindus. Before a person desires to enter the jungle, he bends to the ground, hands folded, and says: 'In the name of Ghazi Saheb', thus believing himself delivered into his protection.

It is believed that crocodiles as well as tigers are at the command of the fakirs who can make them rise or sink at will, shut their mouths and prevent them attacking human beings.

When a boatman is killed by a tiger, his oar is planted blade upwards at the place where he was attacked, and a white flag, with some rice tied to a corner of it, is fixed to the oar. If any person attempts to remove the oar and fails to draw it out of the ground with a single pull, it is believed that he will be killed by a tiger; but nobody generally interferes with these simple memorials to the dead, which are seen on the banks of rivers throughout the Sundarbans.

(O'Malley's *Gazetteer for 24 Parganas*, 1914)

A rumble known as the 'Barisal guns', resembling the note of a cannon, is heard in the Sundarbans during the rainy season. The roar appears to come from the south-east i.e. from the seaboard of Barisal (Bangladesh). (However, the writer has not been able to hear the "Barisal guns" during his visits to the region).

(*O'Malley's Gazetteer for Khulna*, 1914)

The fakirs are easily distinguished in the Sundarbans; they are invariably quick intelligent men with sharp eyes and a wild look about the face. They have great influence; not a man will step out of his boat and enter the jungle, unless preceded by a fakir; not a follower or beater will attend a sportsman in the forest, unless he engages the services of one of these men to secure the tigers away or to shut their mouths. These men do not even pretend to have the least influence over rhinoceros or buffaloes, but only over tigers.

To the sportsman, the Sundarbans during the cold season is a great treat. It is true there are no elephants there to shoot from, and even if there were, they would be useless in the forest; but it is a pleasure to roam the pathless woods, to trust your own skill and daring, and not to be indebted for sport to the skill of your mahout. There we have the royal Bengal tiger in his native home, lord of the forest, with none to dispute with him a fat hog or a fat deer. There is the huge rhinoceros with his single horn, lazily feeding all night on the young branches of the *keora* and wallowing all day in muddy pits. There the spotted deer bound past you at every open glade. At early dawn or sunset, they are to be seen in herds of 20 and 30, feeding at the sea shore, with the dark wood on the side and the open blue sea on the other. There too, in the deep forest where hardly a single ray of light penetrates the mass of foliage over head, suddenly from under your feet starts the barking deer with his tiny tusks and the woods re-echo with his wild and startling bark. In the more open parts, the fat hog-deer bounds before you; there are the monkeys chattering away on the trees; and where the jungle is heavy, outleaps with a strong and mighty bound the great swamp deer with its antlered head. There, in every direction, we see the wild hog at his everlasting work busily digging away at the roots of the old trees. There too, last but not least, is the monster buffalo which is nearly as huge as the rhinoceros. There also in the slime and mud on the banks of river, is seen the great crocodile with his tough, hard skin almost impervious to ball, lazily gliding into the water and disappearing with a splash. These are the principal fauna of the Sundarbans, and right good sport do they afford to a sportsman, who has a stout pair of legs, a good rifle on his shoulder and a compass in his pocket.

The first attempt on the part of the East India Company to cultivate the Sundarbans was made in 1783-90 through Mr. Tilman Henckell, Superintendant of the Sundarbans. His report addressed to Warren Hastings, Governor-General of India, inter alia, stated :

... it is practicable to populate these wild and extensive forests, not a mere speculative idea, we have only to recur to the times of the Mughal Government, and we shall find, that prior to the invasion of the *Mag* in the Bengali year 1128, these lands were in the finest state of cultivation, and the villages in general well populated. The number of mosques and other places of worship still remaining, fully demonstrate its former splendour and magnificence. Nature also has been particularly lavish and bountiful of her favours to this part of Bengal; the number of fine rivers with which it abounds, renders it so convenient for transportation of all kind of merchandise, and its vicinity to Calcutta, the seat of Govenment, affords the merchant and manufacturer a sure prospect of receiving the reward of their labours by the speedy sale of their merchandise, the greatest encouragement to revenue. The quantity of wood and timber proper for constructing boats; the transportation of fire-wood to Calcutta; the quantity of wax that is everywhere found in the woods; and the preparation of shell-lime, will amply reward the ryot for the trouble and expense of clearing away the ground. All that he requires is an assurance of being protected by Government, in the quiet possession of the little spot that he has cleared away by the sweat of his brow.

(*The Calcutta Review*, 1858)

Francois Bernier's description of his journey through the Sundarbans in 1665-66 reads like this:

I remember a nine days' voyage that I made from Pipli (one time port on the Orissa coast) to Ogouli (Hooghly), among these islands and channels, which I cannot omit relating, as no day passed without some extraordinary accident or adventure. When my seven-oared *scallop* had conveyed us out of the river of Pipli, and we had advanced three or four leagues at sea, along the coast, on our way to the islands and channels, we saw the sea covered with fish, apparently large carp, which were pursued by a great number of dolphins. I desired my men to row that way, and perceived that most of them were lying on their side as if they had been dead; some moved slowly along, and others seemed to be struggling and turning about as if stupefied. We caught four-and-twenty with our hands, and observed that out of the mouth of every one issued a bladder, like that of a carp, which was full of air and of a reddish colour at the end. I easily conceived that it was this bladder which prevented the fish from sinking, but could never understand why it thus protruded, unless it were that having been long and closely pursued by the dolphins, they made such violent efforts to escape, that the bladder swelled, became red, and was forced out of the mouth.

The day following we arrived, at rather a late hour, among the islands; and having chosen a spot that appeared free from tigers, we landed and lighted a fire. I ordered a couple of fowls and some of the fish to be dressed, and we made an excellent supper. The fish was delicious. I then re-embarked, and ordered my men to row on till night. There would have been danger in losing our way in the dark among the different channels, and therefore we retired out of a main channel in search of a snug creek where we passed the night, the boat being fastened to a thick branch of a tree, at a prudent distance from the shore. While keeping watch, I witnessed a Phenomenon of Nature such as I had twice observed at 'Delhi'. I beheld a lunar rainbow, and awoke the whole of my company, who all expressed much surprise, especially two Portuguese pilots, whom I had received into the boat at the request of a friend. They declared that they had neither seen nor heard of such a rainbow.

The third day, we lost ourselves among the channels, and I know not how we should have recovered our right course, had we not met with some Portuguese, who were employed in making salt on one of the islands.

In the evening of the fourth day we withdrew, as usual, out of the main channel to a place of security, and passed a most extraordinary night. Not a breath of wind was felt, and the air became so hot and suffocating the we could scarcely breathe. The bushes around us were so full of glow-worms that they seemed ignited; and fires resembling flames arose every moment to the great alarm of our sailors,who did not doubt that they were so many devils.

The night of the fifth day was altogether dreadful and perilous. A storm arose so violent, that although we were, as we thought, in excellent shelter under trees, and our boat carefully fastened, yet our cable was broken, and we should have been driven into the main channel, there inevitably to perish, if I and my two Portuguese had not, by a sudden and spontaneous movement, entwined our arms round the branches of trees, which we held tightly for the space of two hours, while the tempest was raging with unabated force. No assistance was to be expected from my Indian boatmen, whose fears completely overcame them. Our situation while clinging for our lives to the trees was indeed most painful; the rain fell as if poured into the boat from buckets, and the lightning and thunder were so vivid and loud, and so near our heads, that we despaired of surviving this horrible night.

Nothing, however, could be more pleasant than the remainder of the voyage. We arrived at Ogouly on the ninth day, and my eyes seemed never sated with gazing on the delightful country through which we passed. My trunk, however, and all my wearing-apparel were wet, the poultry dead, the fish spoilt, and the whole of my biscuits soaked with rain.

Bernier's *Travels in the Mogul Empire*, Translation, O.U.P., 1914

Only the wealthy classes live in brick houses; the shopkeepers and the husbandmen generally in mud huts. The building materials of a shopkeeper's house consist of bamboo, timber posts, and thatching grass or *golepata* leaves, with mud walls. The number of rooms or huts to each household varies according to the condition of the family. A shopkeeper with a mother, wife and three children would have a hut with two or three verandahs for the dwelling of himself, wife and children; and another hut, to serve both as a cook house and as the dwelling of his mother. A verandah is set aside, or sometimes a separate hut is built, for the purpose of receiving visitors and friends. The dwelling of an ordinary peasant, with the same sized household, would consist of a hut to dwell in, another small one for cooking in, and a cowshed.

Hunter's *Statistical Account of Bengal*, 1874

The most disastrous cyclone within living memory is that of 1864. The storm, which had been slowly travelling up the Bay of Bengal, made itself felt at the Sandheads on the afternoon of October 4 and attained its full fury at night. At Calcutta it raged from 10 a.m. till 4 p.m. on the 5th, after which it gradually subsided; here the lowest reading of the barometer was 28.571 at 2-45 p.m. The destruction caused by the cyclone was twofold. First, the violence of the wind caused widespread destruction to houses and trees. Secondly, the storm-wave brought up by the gale swept over the country to a distance of 8 miles inland on either side of the Hooghly as far north as Achipur. This wave rose in some places to a height of 30 feet, sweeping over the strongest embankments flooding the crops with salt-water and carrying away entire villages. At Sagar Island it was 15 feet above land level, and appeared to cut a channel straight across the island, dividing at into two halves. The embankments, houses, huts, *golas* and buildings were destroyed; and, out of a population of nearly 6,000, less than 1,500 survived. Those did not escape were saved by climbing up trees, or floating on the roofs of their houses, which the wave swept away and carried many miles inland. At Diamond Harbour the wave was 11 feet high, and it was stated at the time that it was impossible to go 50 yards on the road, at any place within six miles of Diamond Harbour, without seeing a corpse. Other villages on either side of the river suffered more or less : in some, every house was swept away with most of the inhabitants. The distress and suffering to which the survivors in the affected tracts were exposed after the disaster were very great. For several days food was not available because the local stores had been swept away , and relief could not be sent from Calcutta. In some places which escaped the storm-wave the stores of the rice merchants were broken open and plundered; in others a kind of grass was eaten as food.

The cyclone wrought havoc among the shipping in the river. On October 5 there were 195 vessels within the limits of the Calcutta Port. They withstood the force of the wind with success; but when to this, at about 1 p.m., was added the storm-wave the force of which was still not entirely spent, one vessel after another broke from her moorings, and as each ship was swept on, she fouled others in her course. Massed together in hopeless and inextricable confusion, they were driven in heaps on the Sumatra Sand and along the Howrah shore from Sibpur to Ghoosery: there was, it must be remembered, no bridge between Calcutta and Howrah in 1864. Ten vessels were sunk in the river and 145 driven on shore. At 6 p.m. the Strand Road was flooded throughout, and in places the water stood breast high. The avenues in Fort William and the Botanic Garden were destroyed: the Eden Gardens were turned into a wilderness: the Barrackpore Park lost 50 per cent of its valuable trees and the avenue on the Barrackpore road suffered even more.

(O'Malley's *Gazatteer*, 1914)

15. Notable Places

Canning — Sagar Island—Fraserganj — Gosaba — Hingalganj — Basirhat —Budge Budge — Diamond Harbour — Falta — Hasnabad

Canning

A town in the Sadar sub-division, situated on the Matla river, it is the terminus of a branch of the railway starting from Sealdah in Calcutta, 45 kms. away. It has a police station, a sub-registry office, and a post office and is also the headquarters of the Sundarbans Tiger Reserve.

The place is named after Lord Canning during whose viceroyalty an attempt was made to establish a port here. In view of the deterioration of the Hooghly, the Chamber of Commerce in 1853 suggested to the Government that a subsidiary port be made at Canning. The Government spent Rs. 11,000 on buying Lot No. 54, with an area of about 8,260 acres, for the purpose of constructing a ship canal and railway to connect the Matla with the Hooghly. The establishment of Port Canning began about 1858. In 1862 the Port Canning Municipality was formed and formally obtained from the Government its right to the town lands. Attempts were made to raise public loans for the improvement of the town and port, but they failed. In connection with this scheme a company was started, called the 'Port Canning Land Investment, Reclamation and Dock Company Limited' for the purpose of purchasing and reclaiming the waste-lands on the river Matla. A railway was constructed between Calcutta and Port Canning, and wharves were built in connection with the railway; but the port failed to attract trade.

Sagar Island

This lies in the Diamond Harbour sub-division and is situated at the mouth of the Hooghly. It is bounded by the Hooghly on the west and by the Channel creek on the east while the Bay of Bengal washes its southern face. The West Bengal Government has a proposal to create more tourist facilities in the island to attract year-round tourism. A proposal to set up a marine park at Sagar-Sandhead is under consideration.

The southern seaface is the site of the great bathing festival of Ganga Sagar on auspicious *Makara Sankranti* day. Every year, about 500,000 pilgrims throng the place from all over the country to bathe in the sea in the early morning. Some have their heads shaved and many of those whose parents have died recently observe the *shardhh* or obsequial ceremonies on the sea-shore. After their ablutions the pilgrims gather at the temple dedicated to Kapilmuni. The image of Kapila is made of stone and is painted red.

O'Malley's gazetteer has recorded the legend behind the sanctity attached to the Ganga Sagar : Sagar, king of Oudh, the thirteenth ancestor of Rama, had performed the *Aswamedha Jajna* (horse-sacrifice), ninety–nine times. This ceremony consisted of sending a horse round the Indian world, with a defiance to all the earth to arrest its progress. If the horse returned unopposed, it was understood to be an acquiescence in the supremacy of the challenger, and the animal was then solemnly sacrificed to the gods. When Sagar made preparations for the hundredth sacrifice, Indra, king of Heaven, who had himself performed the ceremony a hundred times, jealous of being displaced by this new rival, stole the horse, and concealed it in a subterranean cell, where the sage Kapila, or Kapilmuni, was absorbed in meditation, oblivious to all the happenings of the external world. The 60,000 sons of Sagar traced the horse to his hiding-place, and, believing the sage to be the author of the theft, assaulted him. The holy man being thus roused, opened his eyes and cursed his assailants who were immediately burnt to ashes and sentenced to hell. At last a grandson of Sagar, in search of his father and uncles, came to Kapilmuni, and begged him to redeem the souls of the dead. The holy man replied that this could only be effected if the waters of Ganga could be brought to the spot to touch the ashes.

Ganga was residing in Heaven, in the custody of Brahma the Creator. The grandson of Sagar prayed to Brahma to send the goddess to the earth. He died, however, without his prayer having been granted. He left no issue; but a son, Bhagirath, was miraculously born of his widow, and through his prayers Brahma allowed Ganga to visit the earth. Bhagirath led the way as far as

Hathiagarh, in the 24-Parganas near the sea, and then declared that he could not show the rest of the way. Whereupon Ganga, in order to make sure of reaching the spot, divided herself into a hundred mouths, thus forming the delta of the Ganga. One of these mouths reached the cell, and, by washing the ashes, completed atonement for the offence of the sons of Sagar, and their souls were then admitted into heaven. Ganga thus became the sacred stream of the hundred mouths. The people say that the sea took its name of *sagar* from this legend.

The reclamation of the island from jungle was started in the early nineteenth century. In 1811 an European asked for permission to set up a factory for making buff-leather on a plot of hundred acres and requested that all tiger-skins brought to the collector's office might be made over to him. His application was granted, but the project did not get off the ground. The island was subsequently leased to an association of Europeans and Indians, free of rent for the first thirty years and to pay 4 annas per *bigha* thereafter. However, many unforeseen difficulties occurred and till 1820, not more than four square miles had been effectively cleared. It was found that as the woods were cut down, the sea encroached on the land as the sandy beaches failed to resist its invasion. Twenty-five families of *Mags* from Arakan were settled at the confluence of two creeks, and a road constructed for the accommodation of pilgrims to the temple of Kapila.

In 1819, Trower, Collector of the 24-Parganas, originated a company called the Sagar Island Society, for the systematic reclamation and development of the island. The Company carried on operations vigorously until 1833, when their work was destroyed by a cyclone and they abandoned the project. Their interest in the northern part of the island was then taken over by four European gentlemen, who combined the manufacture of salt with the cultivation of rice. The progress of the island was again interrupted by the cyclone of 1864. Now, except for the *mela* (fair) ground, the entire island is inhabited.

Fraserganj

The island is situated in the extreme south of the Diamond Harbour sub–division. It is bounded on the north and west by the Pattibunia Khal, on the east by the Saptamukhi river and on the south, its sandy beaches face the Bay of Bengal. It was called

Fraserganj after Sir Andrew Fraser, Lieutenant Governor of Bengal from 1903 to 1908, during whose tenure a scheme for reclamation and colonization was undertaken. Steps were taken to develop the place as a health resort for the residents of Calcutta. The work of reclamation proved so costly that it was given up. While it was in progress, a number of house sites were discovered surrounded by large tamarind and *mansa* (*Euphorbia nivedia*) trees and in the south-east of the island four old kilns and scattered bricks were found, all going to prove that the island was formerly inhabitated.

The West Bengal Tourism department has a fairly large tourist lodge at Bakkhali, near Fraserganj.

Gosaba

Gosaba, 90 kms. to the south-east of Calcutta, is the headquarters of a thana and a Block. It was the centre of the activities of Sir Daniel Hamilton in promoting rural development through the cooperative movement. Sir Daniel was a shareholder of a British firm in Calcutta. The area of his estate was 22,000 acres, of which 17, 000 acres were under actual cultivation. In accordance with the prevalent system, Sir Daniel began to advance loans to his tenants on easy terms, but found the system unrewarding. He established the first cooperative credit society with 15 members in 1916 at Gosaba. The credit movement gradually spread and new societies financed by him were formed. In 1924 these societies were federated under the Gosaba Central Bank. In 1919 a cooperative store was formed to supply the members with their daily necessities. A Cooperative Paddy Sale Society was established in 1922. The society had boats to transport paddy to Ultadanga, the most imortant paddy mart of Calcutta. The boats went with paddy and returned with purchases for the cooperative store. Sir Daniel set up the Jamini Rice Mill in 1927 as an annexe of the sale society. The mill was named after Mr. J.N. Mitra, the then Registrar of Cooperative Societies, Bengal, as a tribute to the services rendered by him to the movement.

There is an inspection bungalow at Gosaba.

Hingalganj

This village lies on the west bank of the Kalindi river in the Basirhat sub-division. It is one of the main markets for the *abads* (cultivated clearings) in the Sundarbans, where people bring their produce, such as rice, wood and fish, for sale and pick up provisions like salt, kerosene etc.. The place was named after Mr Henckell, Magistrate of Jessore who was appointed Superintendent for the Sundarbans in 1784. Henckell established three markets for the development of the region, one of them being Henckellganj (subsequently distorted to Hingalganj). When his overseer was clearing the forest, the work was interrupted by tigers. The overseer therefore, called the place Henckellganj in the belief that the tigers would be overawed by the name and cease to molest his men.

Basirhat

A sub-divisional town, it is situated on the right bank of the Ichamati. The town has one building of archaeological interest—the mosque known as Salik Mosque. It is popularly believed that the mosque was built by Ala-ud-din in 1305, but an Arabic inscription over the central gate shows that it was erected by one Ulugh Majlis-i-Azam in 1466–67.

Budge Budge

A town in the Sadar subdivision, it is situated on the bank of the Hooghly. It is the terminus of a branch line of the Eastern Railway.

Budge Budge formerly had a fort which was captured by Clive in his advance on Calcutta in December, 1756. The fort ceased to exist in 1793. The drainage of the town is a difficult problem and most of the buildings are on artificially raised ground. There are several big jute mills here. It is the oil depot of Calcutta where tank steamers discharge.

Diamond Harbour

This is a sub-divisional town, situated on the east bank of the Hooghly, which is here joined by Diamond Harbour Khal. It is 50 kms. south of Calcutta by road and is connected with that city by a branch line of the Eastern Railway. It is the site of the old Chingrikhali Fort were heavy guns were mounted. In the old days, the Company's ships anchored here. An old European cemetery is located in the town. The town suffered severely from the cyclone of 1864 which swept away the majority of its inhabitants. Diamond Harbour is a favourite haunt of Calcuttans during weekends. The West Bengal Tourism Department maintains a fairly large tourist lodge here.

Falta

This is a village in the Diamond Harbour sub-division, situated on the bank of the Hooghly. It is the headquarters of a police station and the site of an old fort. A free trade zone has recently been set up here. In the eighteenth century the Dutch maintained a station here, to which the English retired after the capture of Calcutta by Siraj-ud-daula in 1756 and where they remained until a sufficient force had been collected for its recapture. In the early part of the nineteenth century there was a large agricultural farm here under European management.

Hasnabad

A village in the west of the Basirhat sub-division, situated on the west bank of the Ichamati. It is the headquarters of a police station and a centre of trade on the Sundarbans boat-route. The famous Baptist missionary, Dr. Carey lived here for sometime. Carey wanted to live in a place which afforded him the opportunity of accommodating his habits to those of the Indian community. When he was living at Manicktala in Calcutta, his wife went insane and two of his four children were down with dysentery. His *munshi* suggested that the destitute family should move to the waste jungles of the Sundarbans and cultivate a grant of land there. A boat was hired and on the fourth day, when only one more meal remained, the miserable family and its stout-hearted head saw an English-built house. As they walked up to it, its owner, Charles Short, who was in charge of the company's salt manufacturing centre there, met them and with Anglo-Indian hospitality invited them to be his guests. Carey took a few acres of land and built a bamboo-house. It is said that the local people squatted around Carey's house believing that the sahib's gun would keep away tigers.

16. Tourist Facilities

Communications — Sajnekhali Tourist Lodge — Watch - towers

The only transport system in the Sundarbans is the waterways. In addition to the launch services from Namkhana, Canning, Sonakhali, Nazat and Dhamakhali, most of the inland traffic is carried by country boats with sails. Where the water is shallow, they are poled; but generally they travel with the tides, going up one river with the flood-tide, down another with the ebb and anchoring when the tide is against them. A *pansi* is a broad-bottomed boat with a thatched cabin and a deck of bamboo. The dinghy, used for passenger traffic, is 30 feet long and four feet broad. It has an arched roof of matting in the middle affording protection against weather. These days motorised dinghies are quite common in the region. The *donga*, or dug-out, is used on smaller streams and swamps.

A drive from Calcutta to the embarkation points, Canning (54 km. away) and Sonakhali (100 km. away) takes $1^1/_2$ and $2^1/_2$ hours respectively. At either of these two ports a visitor has to board a motor launch for his journey to the forests.

The 60-bed tourist lodge at Sanjekhali, Sunder Chital (named after the beautiful axis deer), provides comfortable accommodation for visitors. The wooden lodge on stilts looks out on a mangrove forest bordered by a fresh-water tank. Relaxing in the lodge one can enjoy the spectacle of wild animals such as deer, pigs and if luck favours, even a tiger drinking water from the tank. A mangrove interpretation centre has recently been opened near the lodge. Watch-towers are located at Sajnekhali, Sudhanyakhali Netidhopani, Burirdabri, Jhingakhali and Haldi. Sudhanyakhali is hardly an hour's journey by boat from Sajnekhali.

Apart from tours conducted by the West Bengal Tourism aboard the *Madhukar*, visitors can reach Gosaba by the public launch service from Canning and proceed to Sajnekhali by motor boat or van rickshaw.

The region is visited by storms blowing from the north-west in the evening during April and May. These are locally called *kal baisakhi* (nor'westers). Though these are of short duration they are severe enough to cause harm to unwary boatmen, standing crops, fruit trees and houses. Cyclones of the equinox, occurring in March and September, are also major storms in the region. The best season for visiting this natural haven, the Sundarbans, is winter, between the months of November and February. The future will bring an ever-increasing number of visitors to the enchanting forests of the Sundarbans.

KEY WORDS TO APPENDIX I

T	=	Tree
S	=	Shrub
C	=	Climber
H	=	Herb
Gr	=	Grass
ME	=	Mangrove Exclusive

Appendix — I

LIST OF PLANTS IN THE SUNDARBANS TIGER RESERVE

	Scientific Name	Local Name	Family
1.	*Acanthus ilicifolius L. (ME)*	Harguza	Acanthaceae (S)
2.	*A. volubilis wall (ME)*	Lata harguza	Acanthaceae (C)
3.	*Aegialitis rotundifolia Roxb (ME)*	Tora	Plumbagenaceae (S)
4.	*Aegiceras carniculatum (L) Blanco. (ME)*	Khalsi	Myrsinaceae (T)
5.	*Avicennia alba Blume (ME)*	Piara baen	Verbenaceae (T)
6.	*A. marina (Forsk) Vierh.(ME)*	Kala baen	Verbenaceae (T)
7.	*A. officinalis L. (ME)*	Jat baen	Verbenaceae (T)
8.	*Aglaia cuculata (Roxb) Pellegrin.*	Amur	Meliaceae (T)
9.	*Acrostichum aureum L.*	Hodo	Pteridaceae (S)
10.	*Bruguiera cylindrica (L) Blume (ME)*	Bakul	Rhizophoraceae (T)
11.	*B. gymnorhiza (L) Lam. (ME)*	Kankra	Rhizophoraceae (T)
12.	*B. Sexangula (Lour) Poirot.(ME)*	Kankra	Rhizophoraceae (T)
13.	*B. parviflora (Roxb) Wight and Arn. (ME)*	Bakul	Rhizoporaceae (T)
14.	*Brownlowia tersa (L) Koster.*	Lata-sundari	Tiliaceae (S)
15.	*Cerbera manghas Geartn*	Dakor	Appocynaceae (T)
16.	*Clerodendrum inerme (L) Geartn.*	Banjai	Verbenaceae (S)
17.	*Caesalpinia bunducella (L) Roxb.*	Nate	Leguminoseae (S)
18.	*C. cristata (L)*	Nate	Leguminoseae (S)
19.	*Ceriops decandra (Gritt) Ding Hou. (ME)*	Jhamti or jele goran	Rhizophoraceae (S)
20.	*C. Tagal (Perrottet) C. S. Robinson (ME)*	Math goran	Rhizophoraceae (S)
21.	*Cynometra iripa Kostel.*	Singra	Leguminoseae (T)
22.	*Derris scandens (Roxb) Benth.*	Kalilata	Leguminoseae (C)
23.	*D. heterophylla (Willd) Back and Bakh.*	Kalilata	Leguminoseae (C)
24.	*Dalbergia spinosa Roxb.*	—	Leguminoseae (C)
25.	*Entada scandens Benth.*	Gila	Leguminoseae (C)
26.	*Excoecaria agollocha L. (ME)*	Genwa	Euphorbiaceae (T)
27.	*Finlaysonia obovata L.*	Dudhilata	Asclapiadeceae (C)
28.	*Fimbristylis ferruginea*	—	Cyperaceae (H)
29.	*Hemithrea compressus L.*	—	Graminae (Gr.)

	Scientific Name	Local Name	Family
30.	*Heliotropium curassavieum Linn.*	—	Chenopodiaceae (H)
31.	*Heritiera fomes Buch-Ham.(ME)*	Sundari	Sterculiaceae (T)
32.	*Hibiscus tilliaceous L.*	Bhola	Malvaceae (T)
33.	*Kandelia candel (L) Druce (ME).*	Garia	Rhizophoraceae (T)
34.	*Lumnitzera racemosa Willd. (ME)*	Kripa	Combrataceae (T)
35.	*Mukuna gigancea (Willd) D. C.*	—	Cucorbitaceae (C)
36.	*Myriostachya wightiana (Nessex Steud) Hookf.*	Ghash	Graminae (Gr.)
37.	*Nypa fruiticans Wurmb. (ME)*	Golpata	Palmae (T)
38.	*Phoenix paludosa Roxb. (ME)*	Hental	Palmae (T)
39.	*Porterasia coarctata (Roxb) Tateoca. (ME)*	— Dhanighash	Poeceae (Gr.)
40.	*Pandanus tectorius soland*	Keya	Pandanaceae (T)
41.	*Pluchea indica Linn.*	—	—
42.	*Rhizophora apiculata Blume.(ME)*	Garjan	Rhizophoraceae (T)
43.	*R. mucronata Lam. (ME)*	Garjan	Rhizophoraceae (T)
44.	*Sacharum cylindricum L.*	Ulu	Graminae (Gl.)
45.	*Suaeda nudiflora Mog.*	Gira sak	Chenopodiaceae (H)
46.	*S. monoeca L.*	Nonaguru	Cenopodiaceae (H)
47.	*S. meritima Dumort.*	Nonaguri	Chenopodiaceae (H)
48.	*Sesuvium portulicastrum L.*	Jadu Palang	Ficoidae (H)
49.	*Salicornea brachiata Roxb.*	—	Chenopodiaceae (H)
50.	*Salacia prinoides (Willd) D. C.*	—	Celastraceae
51.	*Scirpus littorea Schrad.*	—	Cyperaceae (H)
52.	*Sonneratia apetala Buch-Ham.(ME)*	Keora	Lythraceae (T)
53.	*S. caseolaris (L) Engl. (ME)*	Ora	Lythraceae (T)
54.	*S. griffithii Kurz. (ME)*	Ora	Lythraceae (T)
55.	*S. alba J. Smith (ME)*	Ora	Lythraceae (T)
56.	*Stenochlaena palustre Bold.*	—	—
57.	*Stichtocardia tilifolia Hallier F.*	—	Manispermaceae (C)
58.	*Thespesia populnoides (Roxb) Kostel.*	Paras	Malvaceae (T)
59.	*T. populnea (Linn) Solex. Carrea.*	Paras	Malvaceae (T)
60.	*Tamarix troupii Hole.*	Nona Jhau	Tamaricaceae (T)
61.	*T. dioica Roxb.*	Nona Jhau	Tamaricaceae (T)
62.	*Viscum orientale Van.*	—	Loranthaceae (C)
63.	*Xylocarpus granatum Koenig.(ME)*	Dhundul	Meliaceae (T)
64.	*X. mekongensis Pierre. (Sym. X. gangeticus Parkinson)(ME)*	Passur	Meliaceae (T)

Appendix II

LIST OF IMPORTANT ANIMALS OF THE SUNDARBANS TIGER RESERVE.

	Scientific name	English/local name
1.	*Panthera tigris*	Tiger
2.	*Felis viverina*	Fishing cat
3.	*Felis chaus*	Jungle cat
4.	*Viverra zibatha*	Civet cat
5.	*Viverricula indica*	Small Indian civet
6.	*Cervus axis*	Axis deer or chital
7.	*Sus scrofa*	Wild boar
8.	*Macaca mulatta*	Rhesus monkey
9.	*Varanus salvator*	Water monitor or salvator lizard
10.	*Varanus flaviscense*	Yellow monitor or monitor lizard
11.	*Crocodilus porosus*	Estuarine crocodile
12.	*Platanista gangetica*	Gangetic dolphin
13.	*Chilleseyllum griseum*	Tiger shark
14.	*Stegestoma fasciatusu*	Tiger shark
15.	*Ophiophagus hannah*	King cobra
16.	*Naja naja*	Indian cobra
17.	*Bungurus fasciatus*	Banded krait
18.	*Vipera russelli*	Russell's viper
19.	*Praescutata viperina*	Sea-snake
20.	*Enhylrina schistosa*	Sea-snake
21.	*Hydrophis obscurus*	Sea-snake
22.	*Carcinoscorpius rotundicallada*	Horseshoe crab
23.	*Batagur baska*	Common batagur
24.	*Lepidochelys olivacea*	Ridley turtle
25.	*Eretmochelys imbrieata*	Hawksbill turtle
26.	*Lissemys punctata*	Pond turtle
27.	*Hilsa ilisba*	Hilsa
28.	*Hilsa toli*	Hilsa
29.	*Lates calcarifer*	Bhekti
30.	*Liza tada*	*Kalagachhi bhangan*
31.	*Liza microlepis*	*Ram parse*
32.	*Mugil oephalus*	*Aadh bhangan*
33.	*Liza parsia*	*Parse*
34.	*Penaeus monodon*	Prawn
35.	*Boleophthalmus boddarti*	*Menomachh*
36.	*Scylla serrata*	Red crab

	Scientific name	English name
37.	*Coenobita cavipes*	Hermit crab
38.	*Diogenes avarus*	Hermit crab
39.	*Clibararius padavensis*	Hermit crab
40.	*Metaplax crenulata*	Metaplax crab
41.	*Metaplax dentipes*	Metaplax crab
42.	*Metaplax distincta*	Metaplax crab
43.	*Uca acutus*	Fiddler crab
44.	*Uca lactea*	Fiddler crab
45.	*Uca dussumierf*	Fiddler crab
46.	*Uca triangularis*	Fiddler crab
47.	*Sesarma bidens*	Tree-climbing crab
48.	*Sesarma impressa*	Tree-climbing crab
49.	*Sesarma tacniolatum*	Tree-climbing crab
50.	*Sesarma tetragonum*	Tree-climbing crab

Appendix III

LIST OF BIRDS SIGHTED AT SAJNEKHALI AND SUDHANYAKHALI

	Scientific name	English name
1.	*Phalocrocorax niger*	Little cormorant
2.	*Ardea purpurea*	Purple heron
3.	*Butorides striatus*	Little green bittern
4.	*Ardeola grayii*	Pond heron
5.	*Bubulcus ibis*	Cattle egret
6.	*Egretta alba*	Large egret
7.	*Egretta intermedia*	Median egret
8.	*Egretta garzetta*	Little egret
9.	*Nycticorax nycticorax*	Night heron
10.	*Anastomus oscitans*	Openbill stork
11.	*Leptoptilos javanicus*	Smaller adjutant stork
12.	*Haliastur indus*	Brahminy kite
13.	*Vanellus indicus*	Redwattled lapwing
14.	*Numenius phacopus*	Whimbrel
15.*	*Limosa limosa*	Blacktailed godwit
16.*	*Tringa hypoleucos*	Common sandpiper
17.*	*Streptopelia decaocto*	Ring dove
18.	*Streptopelia chinensis*	Spotted dove
19.	*Psittacula krameri*	Roseringed parakeet
20.	*Centropus sinensis*	Crow pheasant
21.	*Cypsiurus parvus*	Palmswift
22.	*Ceryle rudis*	Pied kingfisher
23.	*Halcyon orientalis*	White-collared kingfisher
24.	*Dinopium benghalense*	Golden-backed woodpecker
25.	*Picoides macei*	Fulvus-brested pied woodpecker
26.	*Hirundo rustica*	Common swallow
27.*	*Mirops orientalis*	Small green bee-eater
28.	*Dicrurus aeneus*	Bronzed drongo
29.	*Artamus fuscus*	Ashy swallow-shrike
30.	*Sturnus contra*	Pied myna
31.	*Acridotheres fuscus*	Jungle myna
32.	*Corvus macrorhynchos*	Jungle crow
33.	*Coracina melanoptera*	Blackheaded cuckoo-shrike
34.	*Coracina novaehallandiae*	Large cuckoo-shrike
35.	*Aegithina tiphia*	Common iora
36.	*Pycnonotus cafer*	Redvented bulbul
37.	*Rhipidura albicollis*	Whitespotted fantail flycatcher
38.	*Orthotomus sutorius*	Tailor bird

	Scientific name	English name
39.	*Capsychus saularis*	Magpie robin
40.	*Nectarinia asiatica*	Purple sunbird
41.	*Zosterops palpebrosa*	White-eye
42.	*Passer domesticus*	House sparrow
43.	*Lonchura malacca*	Black-headed munia

* marked birds are migratory, others are local

Index